Seaweeds at Ebb Tide

Seaweeds at Ebb Tide

By
MURIEL LEWIN GUBERLET

Illustrated by Elizabeth L. Curtis

UNIVERSITY OF WASHINGTON PRESS
Seattle and London

Copyright © 1956 by the University of Washington Press
First printing, clothbound edition, 1956
Second printing, first paperback edition, 1961
Third printing, paperback edition, 1966
Fourth printing, paperback edition, 1967
Fifth printing, paperback edition, 1971
Library of Congress Catalog Card Number 56-9308
Printed in the United Sates of America

To
DOROTHY MACDONALD
who
suggested that I
write this book

Nature is part of our humanity and without some awareness of that divine mystery man ceases to be a man. When the Pleiades and the wind in the grass are no longer a part of the human spirit, a part of the very flesh and bone, a man becomes, as it were, a kind of cosmic outlaw, having neither the completeness of an animal nor the birthright of a true humanity.
——Henry Beston, The Outermost House

*As we strolled along,
It was our occupation to observe
Such objects as the waves had tossed ashore,
Tangle, or weed of various hues and forms,
Each on the other heaped, along the line
Of the dry wrack. And in our vacant mood,
Not seldom did we stop at some clear pool
Hewn in the rock, and, wrapt in pleasing trance,
Survey the novel forms that hung the sides,
Or floated on the surface,—too fair
Either to be divided from the place
On which they grew, or to be left alone
To their own beauty.*
——Author unknown

Introduction

If you are already familiar with the seaweeds, you will find nothing new in this book. On the other hand, if you are a beginning student of marine algae, you will, I trust, benefit from this little volume.

To be more specific, this book is written for those who are curious about the seashore, that strange world that lies between the limits of the high and low tides. Certainly the seashore is one of the most fascinating areas on the earth. On this narrow strip of land there are more plants and animals than anywhere else on the earth. Here, too, they live out their lives almost exactly as they did when the world was young, for in the sea change is slow and time is endless.

No matter what your contact with the seashore has been, whether you make your home within the sound of the waves, spend a few weeks at the beach during the summer, or on brief outings wander along the shore watching the incoming and outgoing tides, you have undoubtedly been curious about the algae. Perhaps you have wondered about the masses of seaweed strewn along the sands, watched the great beds of brown kelp swaying in the offshore waters, or heard the pop of air bladders of the *Fucus* as you stepped upon them. Unfortunately you may have slipped on the bright green sea lettuce that carpets the rocks. Almost certainly you have admired the delicate red sea lace that is washed upon the shore. It is possible that you have given only a passing glance to these strange forms.

But I am sure that if you become acquainted with a few of the algae and think of them in relation to the wonder of the sea and the universe, your visits to the seashore will take on added meaning and delight. By picking up this book, you indicate your desire to make the acquaintance of this interesting group of plants, so I shall assume that you are willing to put a little work into the study. You will learn the names, habits, structures, and economic uses of the seaweeds. You may have to get up early in the morning to catch the low tides, get wet and dirty as you scramble over the slippery rocks, feel baffled trying to distinguish the species of algae. Don't let that bother you. Even the highly trained botanists have difficulty in identifying them all.

Even if you are unwilling to subject yourselves to the rigors of seaweed collecting and stroll only for fun on your own beach, you may discover a dozen or twenty species almost any day with only average tides. Our Pacific Coast beaches, especially the rocky ones, are very rich in seaweeds. But I can assure you that the satisfaction you will get from these outings will far outweight the discomforts.

The seaweeds, of course, constitute but a small and lowly part of the plant world, for they are nearly at the bottom of the evolutionary ladder. However, seaweeds are an important link in the evolutionary process since they were among the first true plants to develop on the earth. In spite of their early beginnings and their simple structure, there is great diversity among them. The only things they all have in common, and the only conditions indispensable to their existence, are light, air, and moisture. Yet seaweeds differ widely in cell structure, reproductive processes, size, shape, color, and habitat. For these reasons algae which look alike to the naked eye may belong to entirely different genera, families, and even orders.

The variations in color have long been the basis of classification of algae, so all seaweeds are divided into Browns (Phaeophyta), Reds, (Rhodophyta), Greens (Chlorophyta), and Blue-Greens (Myxophyta). The Browns are typically brown or olive-green; the Greens are grass-green or olive-green; the Reds are rose-red to deep purple-red to almost black. The Blue-Greens are not discussed in this book because most of them are microscopic and are thus outside our sphere of interest. Therefore, following the traditional pattern, I have arranged the species according to color.

Although the algae described in this book were collected in the Pacific Northwest and the range of each species is given for the Pacific Coast, most of the genera and many of the species discussed here can be found along either the Pacific or Atlantic coast of North America, and some are world-wide in distribution.

In order to make this study of the algae easy, I have described only the genera and species that you can identify in the field and have confined the descriptions to those points that can be seen with the naked eye or with a hand lens. The discussion of each species is given in one or two paragraphs with the illustration on the facing page. No doubt the illustrations will be as valuable in identification as the descriptions. In addition to the facts about the seaweed under discussion, I have added a paragraph, entitled "Notes on Sea and Shore," which gives some general information about the sea and seaweeds and the complex relationships existing between the physical and biological phases of the all-encompassing sea. I trust this somewhat miscellaneous information will encourage you to delve into more profound and exact discussions of the algae.

I have used technical terms only when familiar ones could not be found to give the idea I wished to express. Since few of the algae have well-established common names, I was forced to coin names for a number of species. Some of these may seem farfetched, but possibly not more so than "Bloody Mary" for a battlefield in Korea or "Dead Man's Bay" for an inlet on Puget Sound. However, if you study the plants at all scientifically, you will be obliged to learn the scientific names which are recognized the world over.

Of course, there are many more seaweeds in Pacific Northwest waters than those discussed in this book. Many species live so deep in the water that they must be dredged; others are seen on the shore only when the tides are extraordinarily low; some are too small to be distinguished without a microscope; and to the unaided eye some species are so similar to other species that it seemed best not to attempt to differentiate them. I do not even claim that all the common, easily distinguishable species have been considered. It is practically impossible to be sure that some typical species have not been missed or that some comparatively rare species have not been included. But in order to discover the general seasonal seaweed inhabitants, Miss Curtis and I have spent several summers (and winters, too, when we could endure the cold) scouring the beaches for representative species.

Hitherto there has not been a simple, nontechnical book dealing with the marine plants of the North Pacific Coast. There are several ponderous and highly technical studies of the seaweeds in the area, but these do not serve the needs of the beginning student or the seaside rambler; therefore I hope this elementary study will fill a need and serve a useful purpose.

It is impossible to thank specifically the many persons who helped and encouraged me in this study. First, I wish to express my sincere gratitude to Dr. Richard Norris, an algologist at the University of Minnesota, who identified most of the species and who read and constructively criticized the manuscript. He was assisted in the identification by Dr. Robert F. Scagel of the University of British Columbia. Thanks are also due to Dr. George B. Rigg, Professor Emeritus of Botany, University of Washington, for giving the manuscript its first critical reading and to Dr. C. Leo Hitchcock, Professor of Botany, for the use of the University of Washington Herbarium and for plant presses. I gratefully acknowledge my debt to Dr. Gilbert M. Smith for the information taken from his book, *Marine*

Algae of the Monterey Peninsula. I used it constantly, especially in indicating the distribution and habitat of the species. My special thanks go to Miss Dorothy Macdonald for suggesting the writing of this book, and for offering Miss Curtis and myself the use of her summer home on Whidbey Island, Washington, as a base of operations, and for refreshing us with hot coffee and shower baths after each collecting foray.

M. L. G.

Collecting Algae

Before algae can be studied they must be collected. Therefore, a few hints about the methods of collection should prove helpful.

The algae discussed in this book are mostly littoral—that is, inhabiting the zone between the high- and low-water marks—and only those species that can be seen with the unaided eye are treated. For the collecting of algae in this zone, you will need very simple equipment. First, you should have a two- or three-gallon pail or canvas bag with a broad strap. To this add a few pint jars and smaller wide-mouth bottles, a tool for scraping or breaking rocks—such as a hammer or heavy knife—and a fairly powerful pocket lens. Sheets of paraffin paper are excellent for wrapping small algae which might become lost among larger species.

In cold regions you will find that hip boots are invaluable although inconvenient. More comfortable are heavy canvas shoes which come above the ankles for protection against barnacles. The soles should be heavy rubber and deeply corrugated. Jeans or slacks and a heavy shirt for protection against the sun and cuts are better than a bathing suit. On a raw, chilly day be sure to wear a windbreaker. But in any event you should expect to get wet since algae grow in the water.

The next prerequisite is a knowledge of the character and time of the tides. It is possible to do some collecting at any time, but of course the greatest variety and number of seaweeds can be found when the tide is low. Tides reach their high and low levels twice each day, and on each succeeding day these extremes occur about fifty minutes later. Extreme high or spring tides come monthly, when the moon and sun are in conjunction at new and full moon, and alternate with minimal or neap tides when the forces are in opposition near the first and third quarters of the moon. Good sources for exact information about the tides are tide charts and local newspapers.

On a collecting trip you walk out over the tide flats or beach as the tide goes out, keeping at the water's edge, examining the pools, rock crevices, and overhanging rocks, and looking under the tangle of rockweeds and the large kelps for small species. It is best to get to the beach about an hour before the water reaches its low mark. Once the tide has turned, keep a sharp lookout for its advance so as not to be caught far out on a boulder or at the foot of a cliff by the rapidly flooding water.

Also be careful when walking over rough stones, for they are often so densely covered with seaweeds that they are very slippery.

The type of bottom largely determines the number and kinds of seaweed you will find. Most sandy beaches or muddy areas yield little seaweed, because on these shores the algae have nothing to attach themselves to. Gravelly beaches likewise harbor little seaweed. Shallow coves and sluggish bays are favorable to certain species of seaweed, but rocky shores provide by far the greatest number and kinds of algae. Sometimes as many as a dozen species may adhere to a single boulder, the green, brown, and red algae forming a riot of color and a profusion of shapes and sizes. Although the tops of submerged rocks offer good collecting, the vertical sides are better, especially if you push away the larger species and work down among the smaller ones.

Only living and healthy mature material with their holdfasts should be collected. Likewise gather more than one individual of a species to show a range in size and shape of blades and branches and other variable features. However, do not be scornful of the material drifting ashore, for often deep-water forms, ordinarily not available without the use of a dredge, are carried in with the returning tide. Also inspect sheltered spots in which algae accumulate, since wind and current tend to localize rich harvests of seaweed.

Only from plants actually attached, growing in their natural habitats, will you learn the habits, distribution, and ecology of seaweeds. As you gather the material, place the small, delicate specimens in pint jars or bottles with salt water, while the larger, coarser specimens are placed in a bucket with a little salt water. It is more important to keep the water cool than to have the algae floating about in a lot of water. The large kelps and rockweeds may be carried in a sack if kept moist. It is better to collect intensively over a limited area than superficially over a large area. Be sure to record the collection data on strong paper with a heavy pencil. Collection data should include the approximate height above low water, the substratum, and the place of collection.

Eventually you will bring the material to your base of operations and you will have to care for and preserve the algae. There are two methods of preservation—in liquid and by drying. For most seaweeds drying is the better method. The equipment for this is quite simple. Coralline algae and rock algae *(Lithothamnium)* can be rinsed with fresh water and dried in a shady place. The kelps and coarser rockweeds can be spread out on the ground and when nearly dry rolled up and folded whole. If these large specimens are to be a part of a permanent collection, you should place them in a mixture of 30 per cent glycerine, 30 per cent alchol, 10 per cent carbolic acid, and 30 per cent water until thoroughly penetrated. When dried they will retain their flexibility for a long time if stored in tin boxes.

Smaller species are best mounted on a medium weight drawing paper, 11.5 x 16.5 inches. Equipment needed for this includes a number of sheets of unbleached muslin or cheesecloth, herbarium blotters (folded newspapers will serve the purpose), a pan large enough to immerse the mounting paper (laundry trays are excellent), a camel's-hair brush, a dissecting needle, and a pair of scissors.

Wash the specimen free of dirt and place it on the paper, allowing it to spread out in its natural shape. Place enough water in the pan to cover the paper and the specimen. If the seaweed is too thick to look attractive on the paper or bunches up, remove some of the branches, being careful not to obscure the characteristic growing habits. Next arrange the specimen on the paper artistically by using the brush and needle. This must be done under water. Now carefully lift the paper from the pan and drain off the excess water. Lay the paper face up on a blotter and cover with cheesecloth. Several layers of blotters, papers, cheesecloth, and specimens are placed one upon another and laid in a plant press or under a moderate weight for drying. If a large number of specimens are being dried, place corrugated paper between the blotters to allow for ventilation.

Within twenty-four hours change the blotters and replace them with a new set. This process should be repeated until the specimens are completely dry. Remove the cloths and arrange the specimens in folders. In most species of seaweed there is enough mucilage to allow them to stick to the mounting paper. For the species with little mucilage, glue can be added to the water in the mounting pan. If you wash and press the cloths, they can be used over and over again. The blotters, too, if thoroughly dried, can be reused indefinitely. Of course, collection data should be recorded on each specimen. This should include the exact locality where the plant was found, the habitat, the date of collection, the collector's name, and the name of the plant if it is known. Many mounted specimens are very beautiful and artistic and can be used for decorative purposes—place cards, note cards, under glass in serving trays, and framed for pictures.

The second method of preservation is in liquid, but this method is not very satisfactory because the fluids usually affect the cell walls rather seriously. Coarse specimens can be preserved in neutral 4 per cent formaldehyde in sea water if they are kept from the light. More delicate ones should be preserved in 30 per cent alcohol in a weak formaldehyde solution. If the material is to be stored indefinitely, 5 per cent glycerine may be added to 85 per cent alcohol to prevent complete drying. For the amateur collector a number of problems arise in the use of liquid preservatives. Therefore, this method is not recommended.

This brief discussion of the collection and preservation of algae in no way covers all the possible methods, but if you follow these suggestions you will have a more or less permanent collection of the macroscopic seaweeds found between the tides.

Seaweeds at Ebb Tide

Enteromorpha compressa
(Linnaeus) Greville

Green confetti

The green confetti, like the other members of the genus, is widely distributed from the Bering Sea to Lower California. It lives rather high on the beach, rarely extending below low-tide mark. It frequently attaches itself to retaining walls along the beach or is found on rocky shores where fresh water escapes from the cliffs.

The distinctive feature of the green confetti is the flattened tubular central thallus (plant body) from which great numbers of lateral thread-like branches arise. The branches are extremely narrow—possibly 1/32- 1/16 inch wide and 4-7 inches long. Occasionally the lateral branches give the alga a tufted appearance. The flattened branches are about the same diameter throughout most of their length, although they may be slightly narrower at the base than at the apex. The green confetti, a transparent yellowish green, has a single layer of cells embedded in a gelatinous substance. It is usually attached to a rock or other algae by a rhizoidal outgrowth from basal cells. However, it is sometimes detached and floating. The holdfast is perennial, but the thallus disintegrates at the end of each season, reappearing the next spring. Like other members of the genus, the green confetti reproduces through alternation of asexual and sexual generations. The reproductive bodies are discharged through an opening in the outer wall of the cell.

NOTES ON SEA AND SHORE

The depth at which an alga grows depends very largely upon the available light and therefore varies with the latitude. In low latitudes light can penetrate to a greater depth than in higher latitudes. In any case, it is unusual to find large quantities of algae at a depth greater than 150 feet. It has been pointed out that the amount of light which reaches plants in broad daylight, when they are growing in 60-70 feet of water, is comparable to clear moonlight. In the littoral zone the amount of light available to the plants depends on their positions relative to the low-water mark and the time of day at which low water occurs.

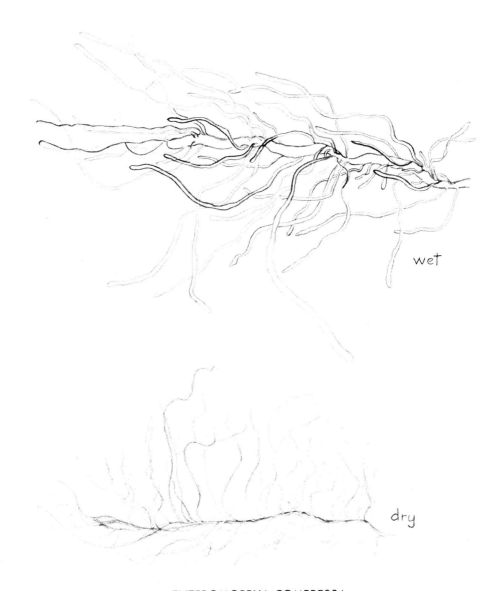

ENTEROMORPHA COMPRESSA

Natural size

Enteromorpha intestinalis (Linnaeus) Link

Link confetti

The most conspicuous characteristic of the Link confetti is the presence of long, tubular, bright green, paper-thin, bladelike branches (thalli). Sometimes there is a single thallus which may be slightly branched near the base, but more often a number of unbranched plants grow together. This species, found on almost every beach from Alaska to Lower California, is attached to rocks or other algae between the 3-foot and the mean low-tide marks. It is held firmly to the substratum by means of rhizoids which are outgrowths of the basal area. When under water the long blades float gracefully on the water; out of water they become matted.

The seaweed has a fragile, tubular structure only one layer in thickness. At intervals the branches are compressed and constricted. When the thallus is deflated, the constrictions are not clearly seen. All the cells except the lowest are capable of producing asexual and sexual bodies which are discharged through openings in the cell wall.

There is a wide variation in the size of this species, the branches sometimes being a yard long and 1/32-1/16 inch wide. However, in some plants the branches are not more than 6 inches long. The Link confetti varies in color from yellow-green to grass-green and is as thin as tissue paper.

NOTES ON SEA AND SHORE

The Link confetti is a common and variable species, almost world-wide in its distribution. There are many intermediate forms which are difficult to separate. This variation is due to differences in habitat, environmental conditions, and the age of the plant. Almost all species of *Enteromorpha* are restricted to salt water; however, certain species live in brackish water and mud flats, while others have moved through the mouths of streams into polluted areas. Some are able to survive in either salt or brackish water with equal ease; a few species attach themselves to the bottoms of boats which travel daily between sea and river ports.

ENTEROMORPHA INTESTINALIS

Natural size

Enteromorpha plumosa Kützing   Silk confetti

The silk confetti looks like a green scum or green silk floating on the water. This unusual plant is found in brackish water all the way from Puget Sound to California. The silky appearance is due to the matting together of the exceedingly fine yellow-green branches. In fact, the branches are so narrow one needs a microscope to study them, and they must be measured in millimeters. The branches are usually from 12 to 24 inches long.

If one takes the trouble to study this seaweed closely, one finds that the branches are cylindrical, somewhat compressed, and much and repeatedly branched. The branches taper and end in a single series of cells. Like the other species of *Enteromorpha,* the green silk has but one layer of cells in the walls of the tubes. The cells are usually embedded in a gelatinous substance and have but one nucleus. The thalli multiply vegetatively by the breaking away of outgrowths. The holdfast is a minute rhizoidal body, but the seaweed is most often found floating on the water.

NOTES ON SEA AND SHORE

Because of the continuous ebb and flow of the tides, the seashore affords two distinct habitats—the part of the shore that is twice daily covered and uncovered by the tides and the region below low tide where the algae always remain submerged. The algae that live in the regions extending between the limits of the high and the low tides must be able to survive periods of exposure of varying duration: drying out, hot sun, and submersion involving changes in temperature, salt concentration, and tide levels. However, some algae live so high on the beach that their vertical range extends above spring tide levels where the only source of salt water is the splash of the tides during rough weather, while those that live far down on the beach are exposed only during the lowest tides. The ability to resist periods of exposure also varies with the genera of seaweed. Some species of *Fucus* and *Porphyra,* for example, can survive out of water 24-48 hours, but the delicate red forms like *Microcladia* can only survive 3-4 hours. The density of the algal population on the rocky shores testifies to the ability of seaweeds to withstand these changing conditions.

ENTEROMORPHA PLUMOSA

Natural size

Ulva lactuca **Linnaeus**   **Sea lettuce**

Sea lettuce is found on the Pacific Coast from Alaska to Lower California attached to rocks or other seaweed from the 2-foot tide mark to the mean low-tide mark. On many beaches of the West Coast, sea lettuce is the most abundant of all algae. It is a grass-green alga with the texture of tissue paper. As it becomes old it may become brown or black.

The sea lettuce has a thin, broadly expanded blade 8-12 inches long with the width approximately $1/4$-$1/3$ the length. The surface of the two-celled blade is perfectly smooth, although the margins are often deeply cut and worn. A number of blades arise from the holdfast so that they resemble a head of loosely arranged leaf lettuce, the branches falling gracefully over the rock to which it is attached. This seaweed lives so high on the beach that it is exposed to the air for several hours at low tide. The holdfast is perennial, the blade annual. The holdfast is formed by outgrowths (rhizoids), produced by the lower cells, which extend downward between the two cell layers and then turn outward to form the holdfast. Sometimes the thalli multiply by breaking apart, producing loose-lying communities. Sea lettuce reproduces through the alternation of sexual and asexual generations. Originally the name *Ulva* was used to include a number of expanded gelatinous algae of any color. Now the name is used in a much more restricted sense.

NOTES ON SEA AND SHORE

Sea water receives nearly 71 per cent of all the radiant energy coming from the sun to the earth's surface. Much of the solar energy is absorbed at or near the surface of the water by minute plants called diatoms. These are almost inconceivably numerous and varied and furnish food materials for microscopic animals. The microscopic plants and animals together are called plankton and constitute the most important manufacturers of food in the ocean. In the English Channel alone 4,000 tons of microscopic vegetable matter are produced annually per square mile.

ULVA LACTUCA

One-half natural size

Ulva linza Linnaeus   Green string lettuce

The green string lettuce grows on rocks or other seaweed from the 2-foot tide level to the average low-tide mark from Alaska to Lower California. This species of *Ulva* is easily recognized by the transparent, bright green color, the long slender blades, the ruffled margins, and the swirled appearance. The blade, from 6 inches to 2 feet in length and from ½ to 1 inch in width, tapers gradually to a short, hollow stipe. The blades, without markings of any kind, are composed of two layers of cells which may be united throughout or may remain free along the margins.

The green string lettuce is interesting because its classification is controversial. The structure of the basal portion is hollow and tubular like the *Enteromorpha*, while the upper portion is flattened and expanded like other species of *Ulva*. It may be considered a link between the two genera. A variety of blades is included under the name *linza* because of differences in length, width, stipe, and margins. Seaweeds, like the *Ulva*, that grow high on the beach must be able to withstand exposure to the hot sun, to periods out of water, and to severe temperature changes. *Ulva linza* is killed by exposure of more than 4-6 hours.

Like the other members of the genus, the green string lettuce is attached by a single cell when young, but later many rootlike outgrowths develop from the lower cells and an attachment disc is formed. This persists through the winter, new plants arising from it in the spring.

NOTES ON SEA AND SHORE

The minute plants and animals in the plankton are consumed by many different species of animals. Certain fish—the herring, for example—devour quantities of plankton by straining it through their gills. In an investigation of the herring in the North Sea some years ago, it was estimated that the herring which were landed on the English coast fishing ports in one year required 109,000 tons of plankton for their food. Within the sea there is a continual struggle for existence. A bigger animal is always waiting to devour a smaller, weaker one; eventually man eats the fish which were nurtured on the plankton.

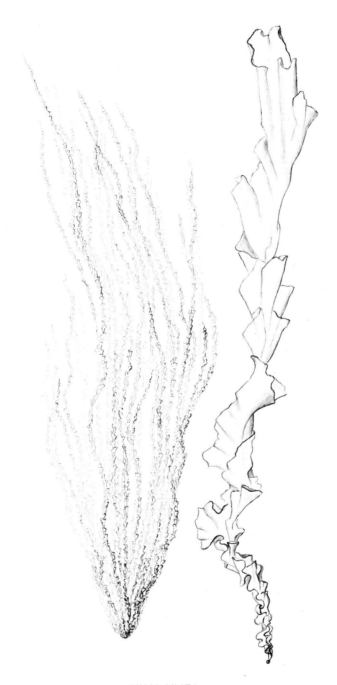

ULVA LINZA

Left, one-sixteenth natural size;
right, one-half natural size

Cladophora trichotoma
(C. Agardh) Kützing

Green ball

The green ball is found from Vancouver Island to Lower California, growing between the 3-foot and the mean low-tide levels. However, it is sometimes difficult to see because it is often almost buried on sand or mud flats.

The most conspicuous features of this seaweed are, as the common name indicates, its bright green color, its ball shape, and the fact that it is buried in the sand. It is about the size of a lime or small lemon. When attached it is ball shaped; when detached it often loses its rounded shape and becomes laterally expanded with a diameter of 1-3 inches. The plant body, made up of profusely branched barrel cells, contains hundreds of tiny branches which divide three or four times, with three to six cells between the branchings. In the center of the body mass is a cavity filled with air. This gives the seaweed its ball-like shape. As the seaweed becomes old it usually loses its ball-like shape and becomes flattened. This species regularly accumulates sand between the branches until every part except the tips is buried. This happens even when it grows on rocks some distance from sandy beaches. The lift cycle includes an alternation of similar generations. The green ball is perennial. It is attached to the substratum by rootlike filaments.

NOTES ON SEA AND SHORE

The species of *Cladophora* are difficult to distinguish. In fact, very few American students have attempted to master the intricacies of the *Cladophora*. The forms on the Pacific Coast are especially difficult to describe accurately. Some writers believe that the characteristic ball shape of *Cladophora trichotoma* is brought about by its being continually rolled over the sand by wave action. One reason for this contention is that the ball forms are found near the shore, while expanded cushion forms are found in deeper water where there is less motion. To refute this theory, it is said that the ball structure seems to be natural because balls have been kept in the laboratory for eight years without losing their shape.

wet

dry

CLADOPHORA TRICHOTOMA

Natural size

Spongomorpha coalita (Ruprecht) Collins   Green rope

The green rope seaweed is abundant along the Pacific Coast from southeastern Alaska to central California. It grows on rocks or other algae at the 1- and the -1.5-foot tide levels.

The distinguishing feature of this seaweed is its bright green, ropelike appearance, the rope being branched and much frayed. The plant body grows 6-8 inches long and ¼ inch wide. The individual filaments or branches are threadlike, but they begin to entangle almost immediately upon arising from the holdfast. The entanglement of the branches is due to special branchlets with two to four recurved prongs projecting from the primary filaments, which hook around one another. When the strands of the rope are separated, one finds that a number of well-defined clusters of filaments grow from a main stem. The cells are joined end to end, which accounts for the very narrow filaments. In the lower part the filaments are forked; in the upper part the branching is alternate or occurs on one side of the filament only. Sand grains often adhere to the rope. This interesting species grows in well-aerated waters. When old the green rope becomes a dull yellowish green. The plant is attached to a rock by a number of branched filaments (rhizoids). After reaching the substratum, the rhizoids divide to form many cells filled with carbonate of lime. The rhizoids persist after the plant dies down at the end of the summer and give rise to new filaments the next year. In other words, the alga has a resting stage.

NOTES ON SEA AND SHORE

Unlike flowering plants, seaweeds do not reproduce by seeds. They form spores of various kinds produced at different seasons of the year. A spore is simpler in structure than a seed, and in its simplest terms can be defined as a cell either with or without the power of spontaneous movement. It eventually becomes free from the parent plant and develops into the new plant. In some genera the development of the spore is extremely complex. In other genera reproduction is very simple, regeneration taking place from the cut surfaces of damaged blades.

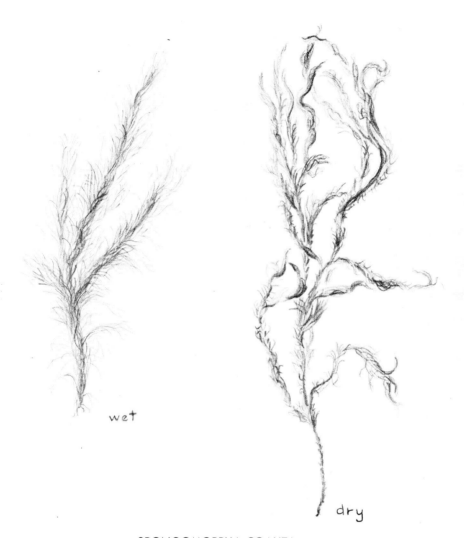

SPONGOMORPHA COALITA

Natural size

Bryopsis corticulans Setchell   Sea fern

From British Columbia to southern California the sea fern grows between the 2- and the -1.5-foot tide levels on rocks exposed to strong surf. Occasionally it is epiphytic on other algae. It is not an abundant species, but one would be sure to notice it if it were in the locality.

The most striking features of the sea fern, a species from 2 to 6 inches in height, are the coarse, stiff branches which extend from a conspicuous axis 1/32-1/16 inch in width, the glossy slippery texture, and the blackish-green color. Just as the primary branches extend stiffly from the axis, so the secondary branches extend stiffly from the primary branches. This gives the plant the regular arrangement of a fern. The lower half of the main axis is bare, while the upper half of the axis has closely crowded branches. Similarly the lower part of the primary branches is bare of branchlets and the upper part closely crowded with them. Both the branches and the branchlets are arranged alternately on their axes. The branches at the top are narrower, more cylindrical, and shorter than the lower ones, giving the whole the shape of a pyramid. Often the seaweed multiplies vegetatively by some of the branches separating from the rest of the thallus. When looked at through a microscope, the cells of this organism are seen to be extremely large. This is an annual plant, fruiting during the spring. The holdfast is a rhizoid from which several stiff, erect shoots arise.

NOTES ON SEA AND SHORE

Because of the constant conditions in the sea, climate does not play so important a part in the distribution of algae as it does in land plants. However, there are powerful factors which tend to limit distribution of algae—great stretches of deep water between localities favorable to seaweed growth, sand banks, strong currents, projecting capes, fresh water from great rivers, and changes in substratum. All these factors tend to prevent the wide distribution of a species whose spores cannot float on the waves beyond a certain period of time without finding suitable places of attachment.

BRYOPSIS CORTICULANS

Natural size

Codium fragile (Suringar) Hariot   Sea staghorn

The sea staghorn grows rather commonly on the tops and sides of rocks between the 1½-foot and the mean low-tide levels from Alaska to Lower California. However, all species of *Codium* are more abundant in warm waters than in cold waters.

The sea staghorn is easy to recognize by the erect, spongy, hairy, forked, cylindrical shoots. The shoots are approximately ⅛- ¼ inch wide and 4-9 inches long. Both the forking and the texture remind one of staghorns. Each of the primary shoots arising from the crustlike base divides into two shoots. These divide again and again with the tips ending in blunt forks. All the branches are approximately the same thickness throughout. The interior of the branches is composed of densely interwoven masses of swollen and thickened filaments. The hairy exterior is covered with mucilage. The spongy shoots are usually a greenish black, but they are sometimes covered with a whitish fleece. The sea staghorn persists throughout the winter, often living for several years.

NOTES ON SEA AND SHORE

The phytoplankton (one-celled floating plants), of course, cannot be seen with the human eye. They are unbelievably small, a thousand of them fitting on the head of a pin. It is said that these minute organisms can reproduce as many as eight times every twenty-four hours, and that one third of the diatoms die every twenty-four hours. When they die they fall to the bottom in quiet waters where they form a reserve of fertilizer until they are circulated and raised to the surface once more. According to E. Frankland Armstrong and L. Mackenzie Miall in *Raw Materials of the Sea,* "The limiting factors of growth in the sea are thus the presence of certain nutrient substances, particularly nitrates, phosphates and perhaps silicates; the presence of various essential trace constituents; and governing both these factors a means of recirculation of vast masses of water so that these materials are used over and over again."

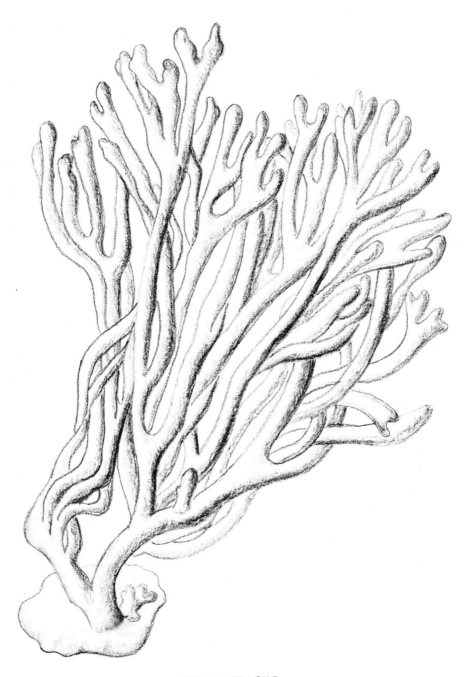

CODIUM FRAGILE

Natural size

Codium setchellii Gardner — Spongy cushion

The spongy cushion grows on exposed faces of rocks between the 1- and the -1.5-foot tide levels. Although it is seldom found in abundance, it appears from Alaska to central California. The less protected the habitat, the more vigorous are the plants.

The striking feature of this seaweed is its unusual shape and texture—a felted, encrusted, dark green or black cushion shaped like a half moon. The flattened body is usually ⅓ inch high and 4-6 inches in circumference. This unusual shape is brought about by a compact covering of non-calcareous pith filaments which end in sacs perpendicular to the surface and held together securely by a gelatinous substance. These filaments bind the entire mass together. The pith filaments are slender, almost colorless threads with one or more air sacs at the upper end and a holdfast at the lower end. The air sacs can be studied only with the aid of a microscope. Young, vigorous specimens are firm and smooth; the old, weak ones are soft and spongy. Sometimes the spongy cushion is attached to the substratum by the central area only, and sometimes it is solidly attached all around the edge.

NOTES ON SEA AND SHORE

The algae of the Pacific Coast are divided into four zones: the Boreal zone extending from northern Alaska to the Strait of Juan de Fuca; the Temperate zone extending from the Strait of Juan de Fuca to Santa Barbara County, California; the Subtropical zone extending from Santa Barbara to Lower California; and the Tropical zone extending from Mexico to the Isthmus of Panama. Most of the species are restricted to one zone, with a small number ranging between two or three zones. A tabulation of the large brown algae showed that 37 of the 49 species reported by Setchell and Gardner are restricted to a single zone; 10 are reported in two zones, and two range through three zones. Pacific shores refute this theory to a certain extent, however, because the Japan Current that skirts a considerable part of the coast produces a pronounced displacement of some species.

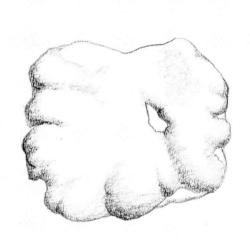

CODIUM SETCHELLII

Natural size

Ralfsia pacifica Hollenberg   Tar spot

The tar spot is common from Alaska to Lower California on rocks which lie rather high on the beach, where it looks like a splotch of very dark brown or black tar dropped on them. Both the size and the shape of the tar spot are indefinite. At first the spot is circular, but it later becomes irregularly lobed, 2-3 inches in diameter, with rounded or scalloped edges. The tar spot is so thin it must be measured in millimeters, a typical specimen being 0.5-1 mm. thick.

With the naked eye the general outline of the seaweed and the radial and circular ridges covering the upper surface can be seen, but a microscope must be used to see the horizontal layers of cells from which arise short cortical threads forming the solid cushion. The cushion is covered with a common layer of very hard mucilage. The microscope shows, too, that clusters of reproductive bodies (sori) nearly cover the surface of the seaweed. Tufts of microscopic hairs on the sori form depressions or warts on the surface. When the reproductive bodies become mature, they break through the mucilage. The fertile threads are also shed, while the parts below continue to grow. Vegetative enlargement, involving both increase in thickness and marginal growth, takes place during the spring and summer. The seaweed is perennial. The tar spot has no specific holdfast, for the entire plant is encrusted on the rock, probably held fast by rhizoids.

NOTES ON SEA AND SHORE

Adjacent tar spots often grow on top of one another so that the structure becomes extremely complex. A person unfamiliar with seaweeds would not dream that the tar spot is an alga. Even when it is looked at through a microscope, its plantlike characteristics are hard to distinguish. However, a number of genera of seaweeds form dark patches on rocks. Likewise several encrusting algae are epiphytic, forming crusts of different sizes, shapes, and colors on the host plant. In these genera, except in the marginal growing region, nearly every cell of the basal stratum produces an erect thread. Most of the upgrowths are short threads of uniform height, closely crowded together and sometimes covered with a common mucilage but not laterally fused.

RALFSIA PACIFICA

Natural size

Heterochordaria abietina   Fir needle
(Ruprecht) Setchell and Gardner

At or just below low-tide mark, the fir needle forms extensive communities on rocks exposed to moderate surf. It ranges from Alaska to southern California.

The striking features of the fir needle seaweed are the lobed, crustlike base and the short, stiff branches surrounding the axis of the shoots. Each shoot has a cylindrical axis with few or many flattened lateral branches extending all around it. The short, rather stiff branches give the seaweed the appearance of a fir tree. The erect shoots are 3-10 inches high and 1/16 inch wide with branches ½-1 inch long. Because a number of plants grow from the same holdfast, they form clusters which look like intricately matted shrubs. Neither the shoot nor the fir needle branches more than once. The axis and branches are solid at first but later become hollow. The growth takes place at the base of the frond. It is believed that the seaweed reproduces with vegetatively identical sexual and asexual spores. The base of the fir needle is apparently perennial, producing several erect shoots each year. The shoots appear in May and disappear in November. When young, the fir needle is light tan, but it becomes dark brown when mature. It has a thick covering of mucilage and is slimy to touch. The holdfast of the fir needle is crustlike and many lobed, sometimes a couple of inches in diameter.

NOTES ON SEA AND SHORE

The fir needle seaweed is known in Japan as matsuma. It is particularly abundant in northern Japan, where it is collected and packed in salt to preserve it for food. Later cooked with a soy bean sauce, it is a common article of food for the poorer people. Another interesting use is in the preservation of mushrooms. The mushrooms are washed in salt water and laid in tight barrels in layers, alternating with layers of matsuma. The people of the Orient use considerable amounts of seaweed as food, but the Japanese are its principal users. In Japan the gathering of the kelp begins in July and ends in October.

HETEROCHORDARIA ABIETINA

Two-thirds natural size

Desmarestia aculeata Lamouroux   Crisp color changer

The crisp color changer is an easily adaptable alga found in the upper littoral zone, in rather deep water, or floating in shallow water from Alaska to Puget Sound.

The most noticeable feature of this alga is the habit of turning bluish green when removed from the water and of discoloring other plants coming into contact with it. When young the alga is light brown; when mature, a dark brown. It has rather harsh, bare branches growing on one plane with a distinct main axis which extends throughout the length of the plant. Both the axis and the branches are decidedly compressed. The alga may reach a length of 6 feet, with the alternate branches arising at intervals of $1/8$-$1/4$ inch. The primary branches are 6-8 inches long; the secondary ones 2-3 inches, hairlike, and widely spaced. At the point where the branches appear there are two shoots, only one of which develops into a branch. When the plant is young, the margins of the branches are covered with fine hairs, but these fall off later. Growth of the axis takes place at the apex, of the branches at the base of the filaments. In *Desmarestia* a single egg extrudes from, but remains attached to, the female sex organ (the oogonium). Fertilization and cell division take place in this exposed position. Later the young plant is dropped into the water. Older portions of the plant body persist through the winter, and new shoots develop from them the next spring.

NOTES ON SEA AND SHORE

The crisp color changer contains phosphates varying from .2-1 per cent of its dry weight. The bluish color taken on after it dies is thought to be due to the action of an acid sap on the fucoxanthin in the pigment cells. Harald Kylin found these aqueous extracts to be acids with a pH of 4.6-6.8. He also concludes that the acid is acetic acid, the fresh alga being as sour as a lemon. However, George B. Rigg believes that the bulk of the acid is sulphuric acid. The bleaching quality of *Desmarestia* is so strong that it extracts color from other seaweeds in contact with it.

DESMARESTIA ACULEATA

Two-thirds natural size

Desmarestia intermedia
Postels and Ruprecht

Loose color changer

Although the loose color changer grows commonly on rocks below tide mark from Alaska to Puget Sound, it prefers deep water in cold latitudes. When torn loose from its substratum, it floats in shallow water on sandy beaches.

The loose color changer has bare, untidily arranged branches borne on an axis extending the length of the plant. Axis and branches lie on one plane. The branching is relatively sparse, with about ¼ inch between successive branchings. Young plants may be 8-10 inches long, while mature ones may be 2-4 feet. The branches of all orders are borne on axes, and are bare, narrow, cylindrical, and tough. Those of the third and fourth degrees are slightly flattened. Young plants are light brown; older ones are dark brown. Like other species of *Desmarestia,* the loose color changer loses its brown color and turns green when taken from the water.

Each plant has a short stipe (1-2 inches), and each branch a short stipe. The alternate branches divide and redivide until those at the ends are fine threads. Near the lower end two or more branches arise from the same node and on the same side of the frond. The axis of the larger branches gives rise to two shoots, only one of which develops into a long branch. On the margins of the younger branches are tufts of hairs. The older parts of *Desmarestia* live for several years, giving rise to new shoots early in the spring of the next year. This species is attached to the substratum by a disc 1/16 inch in diameter.

NOTES ON SEA AND SHORE

When one collects algae, it is important to gather only fresh, living species and to procure entire plants with their holdfast. However, it is legitimate sometimes to gather broken, dead specimens because during storms algae not often found on the shore may be cast onto the beach in large quantities. In this way interesting species are stranded in the drift of the high tide.

DESMARESTIA INTERMEDIA

Three-fourths natural size

Desmarestia munda **Wide branch color changer**
Setchell and Gardner

The wide branch color changer is found from Puget Sound to southern California, attached usually to rocks lying just below low-tide mark, although it may be found 6-8 feet below ebb-tide level. It is frequently cast ashore in quantities and it attracts the attention of seashore visitors.

The most conspicuous feature of the wide branch color changer is the wide, flat, thin central blade that extends the length of the plant. From this central blade (2-4 inches wide), primary opposite branches of the same width as the central blade arise at intervals of 1-3 inches. Running the length of the central blade is a midrib which in the primary branches is inconspicuous. Secondary branches of the same width and arising in the same order may grow from the primary branches. The plant is variable in size, sometimes 2 feet long and sometimes 8 feet long. Branches of all orders are from 1-4 inches in width and borne on short stipes. The tips of the blades are acute or rounded, with short, toothlike projections pointing upward edging the branches. The short, flattened stipe broadens into the blade. The texture of the branches is thin and soft; when dry they become as crisp as tissue paper. Smaller, younger plants are a yellow-brown; larger specimens are a dark brown. In late summer the smaller branches drop off so that during the winter the plant looks naked. For the size of the plant the holdfast is small, not more than $1/4$-$1/8$ inch in diameter.

NOTES ON SEA AND SHORE

The first botanist to collect algae on the Pacific Coast was Dr. Archibald Menzies, a member of Captain George Vancouver's expedition in 1792-1794. Some of the algae collected by Menzies were taken to Europe and described by Dr. D. Turner. The collection was later given to the Edinburgh Botanical Garden. The first paper on the algae of California was published in 1833 by William Henry Harvey; in fact, it was the first paper published on the algae of North America. It is interesting to note that the algae of the Pacific Coast were studied before those of the more accessible and better settled Atlantic Coast.

DESMARESTIA MUNDA

One-third natural size

Punctaria latifolia Greville   **Brown sieve**

The brown sieve floats on quiet waters or attaches to other seaweed from Alaska to Puget Sound. The striking feature of this alga is the presence of holes that dot the single mature blade.

The thin, flat blade varies in shape; sometimes it is long and narrow, at other times almost as broad as long. Although most blades average from 6-10 inches in length and width, some specimens have blades only 3-4 inches wide. In size, shape, and manner of living the brown sieve resembles the green sea lettuce. It is a fragile, paper-thin seaweed, olive brown when in the water, green when removed from the water. The blades are smooth, without markings of any kind, and taper abruptly into the short stipes. The margins of the blades are moderately ruffled. One or more blades may arise from the minute, disc-shaped holdfast.

When the seaweed is mature, reproductive spores project above the blade, forming small dots on the surface. These dots become so numerous that they form dark patches on the blades. When the spores drop off, holes are left in the blades. The seaweed then looks untidy because the blade tears easily, especially when it is pitted with holes. The blade is made up of 2-7 layers of cells, the inner layers longer than the outer ones. The brown sieve is an annual plant which develops through an alternation of dissimilar generations.

NOTES ON SEA AND SHORE

The large, surf-loving algae—*Laminaria, Nereocystis, Macrocystis*—grow in great beds which support a vast number of animals and other algae. In fact, whole communities of plants and animals live out their little lives on the large kelps. These plant hosts offer many advantages to their seaweed tenants—secure holdfasts, escape from the overcrowding of the seashore, and assurance of well-aereated water. Animal tenants also take shelter in the dense growths of the kelp, feed directly upon the host, or prey upon other animals which live in its long blades. Probably two or three dozen permanent "roomers" live on the ribbon kelp.

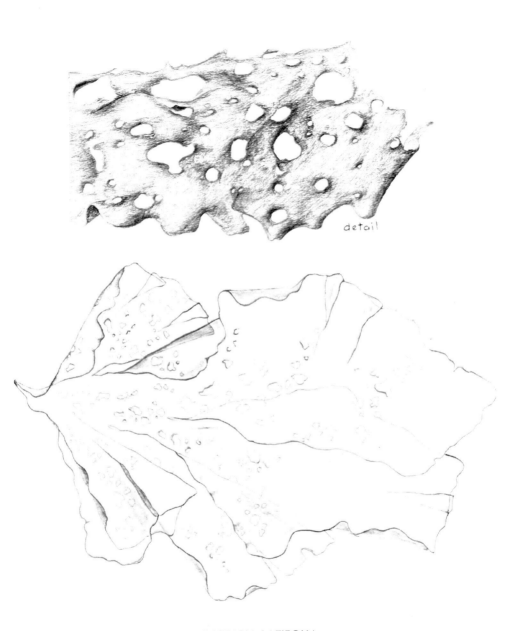

PUNCTARIA LATIFOLIA

Three-quarters natural size

Scytosiphon lomentaria Whip tube
(Lyngbye) J. Agardh

This seaweed grows commonly on exposed rocks or in rock pools from Alaska to southern California. It is partial to the 2.5- to -1.5-foot tide levels. The whip tube is easily recognized by the long, olive-tan to dark brown whiplike shoots which arise from a tiny disc. The flexible, unbranched shoots vary from 8 to 18 inches in length and are 1/16 inch in width. They grow in extensive stands and look like long blades of dry grass.

At first the shoots are solid, but they later become hollow. The hollow end of the shoots may be either constricted (pinched together) or unconstricted, but in either case the shoots are usually twisted in close spirals and gradually taper at base and apex. Smaller individuals showing little or no constriction intermingle with the larger, constricted ones. On the basis of variations in the length of the shoots and the presence or absence of constrictions, the specialists have attempted to describe a number of varieties of this species, but they admit the distinctions are not clear. When the reproductive cells are mature, the fertile areas may cover the entire surface of the plant. Although the alga has many-celled reproductive organs, each cell produces a single reproductive body which is encased in a spore sac. The whip tube is found all over the world.

NOTES ON SEA AND SHORE

The algae, like other pigmented plants, have for centuries been carrying on a process no less amazing than the alchemist's dream of turning lead into gold. This is photosynthesis, by which, with suitable temperatures and light, carbon dioxide reacts with water to form sugar, thus liberating oxygen. Sunlight furnishes the energy to tie together the atoms of the elements present in carbon dioxide and water to produce sugar. The action takes place through chlorophyll, a green pigment made up of carbon, hydrogen, oxygen, nitrogen, and magnesium. We can say that chlorophyll is the most important substance in the world—no chlorophyll, no food; no food, no life.

SCYTOSIPHON LOMENTARIA

Natural size

Colpomenia sinuosa (Roth) Derbés and Solier

Pocket or oyster thief

The oyster thief, a strange alga, is common from Alaska to southern California. It is either epiphytic on another alga or is attached to a mollusc shell at the 0.5- and 1.5-foot tide levels. Occasionally it is dredged from rather deep water. In Puget Sound it frequently attaches itself to *Rhodomela larix*.

The oyster thief is an alga easy to identify. It is an olive-brown balloon, 1½-2 inches in diameter. When young, the seaweed is a globular sac filled with water or gas which acts as a float. It later becomes compressed and has angular indentations. The walls of the seaweed are composed of two layers, an inner layer of large, rounded, colorless cells and an outer layer of small, club-shaped cells. The large sac arises as a result of extensive surface growth by repeated crosswise divisions of the outer layer of cells.

There is considerable variation in the thickness of the walls and the surface texture of the alga. A typical specimen has thin, regular walls, but in some specimens the surface is wrinkled and folded. Older plants may tear open irregularly. In these the surface is covered with blunt warts and spinelike tubercles. The wrinkles resemble the convolutions of a brain. A tuft of microscopic hairs often appears in the depressions. The sex organs first appear around the depressions; later they spread over the entire surface of the plant. The oyster thief is an annual plant. It is attached to the host plant by a broad basal disc made up of a number of threads.

NOTES ON SEA AND SHORE

The name oyster thief is well chosen, for this seaweed is a pest on oyster beds. When the water is shallow or the tide out, the balloon is filled with gas bubbles, and on the return of the tide the inflated balloons lift the young oysters to which they have become attached and float them out to sea. In France workmen have attempted to free the oysters from the seaweed by dragging nets or ropes over the oyster beds. It is believed that the species found on the west coast of the United States was introduced from Europe; it is widely distributed in the warmer seas of the world.

COLPOMENIA SINUOSA

Natural size

Coilodesme californica (Ruprecht) Kjellman — Stick bag

The beginning student of algae may have difficulty in recognizing an epiphytic form (one alga growing upon another but not parasitic upon it), but the stick bag, which lives on *Cystoseira osmundacea* from Puget Sound to southern California, nevertheless seems to merit discussion. The stick bag is abundant at mean low- to -1.5-foot tide levels whenever the host is present.

The stick bag is a long or egg-shaped bag standing at right angles to the branches of the host plant. When young the epiphyte is a smooth, inflated sack or bag, but when it gets old it becomes a thin, flat, eroded blade. In the Puget Sound area the alga is usually 1-3 inches tall and ½-1 inch wide; in California it may become 8-36 inches long and 4 inches wide. In both areas it is an olive-tan color. The bag stands on a very short, solid, cylindrical stipe and is attached to the host by rhizoids. The sack is from 5 to 15 cells thick. These become progressively smaller toward the outer surface. The sporangia (spore cases), which are scattered over the entire bag blade, are completely immersed in the walls of the thallus. The alga appears in May and disappears late in August.

NOTES ON SEA AND SHORE

In what manner the sea produced that wonderful stuff called protoplasm, the essential substance of which life is made, no one knows. It is possible that, after much trial and error, certain organic substances were compounded from carbon, sulphur, phosphorus, potassium, and calcium. These may have been combined to make the complex molecules of protoplasm with the ability to reproduce themselves and begin and endless stream of life. Rachel Carson in *The Sea Around Us* says that these first living things may have been "border line forms, not quite plants, not quite animals, barely over that intangible line that separates the living from the non-living." But it was a beginning.

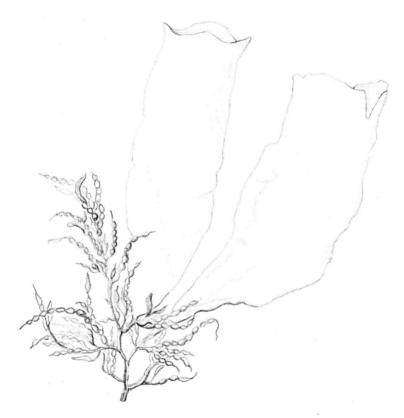

COILODESME CALIFORNICA on CYSTOSEIRA

Natural size

Laminaria platymeris De la Pylaie   Sea girdle or tangle

The sea girdle with its single blade is one of the largest and most conspicuous of the big brown kelps, and some Pacific beaches are dominated by it. This species grows from Alaska to Puget Sound, at or near low-tide mark.

Specimens differ greatly in length, thickness, degree of flattening of the stipe, shape of blade, and splitting of blade. An average specimen may be 3 feet long and 1½ feet wide, with the blade divided twice to a dozen times. The texture of the blade is heavy and leathery and dark brown to black. A single specimen may be so large and heavy that a man can scarcely lift it.

When the plant is young the blade is undivided, but when mature it divides into segments of different widths and lengths. At the base the short, thick stipe (3-7 inches) is cylindrical and tapering, but it broadens into the blade at the upper end. The blade is without markings of any kind, although some plants have a row of blisters a short distance from the margin. Mucilage ducts are present in both stipe and blade, making the surface sticky. The male sex products are in dark patches irregularly embedded in the surface of the blade. The stipe is perennial, but the blade is renewed over a period of years. The holdfast is a large, branched, rounded mass of fibers strong enough to hold the plant in place despite the force of the waves.

NOTES ON SEA AND SHORE

The large brown algae are among the sources of commercial iodine. It is reported that in France 72 tons of iodine were removed from 180,000 tons of seaweed in a single year. In the United States during the First World War, large quantities of seaweed were harvested, pulped, and allowed to ferment, producing acetic acid. After the extraction of calcium acetate, the residue was worked for potash and iodine. However, at the present time little iodine is made from algae in the United States because cheaper and more readily accessible sources are available.

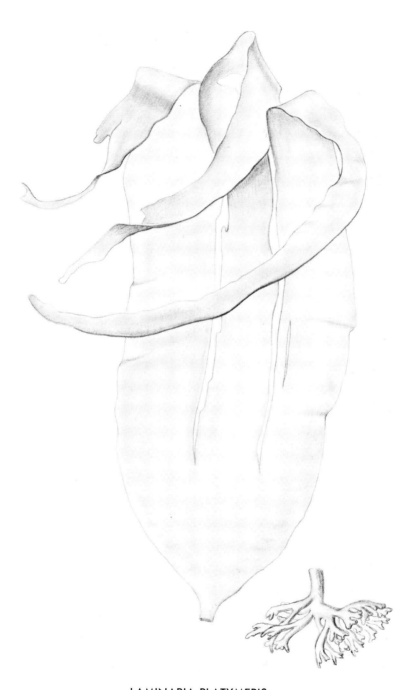

LAMINARIA PLATYMERIS

One-third natural size

Laminaria saccharina (Linnaeus) Lamouroux   Sugar wrack

The sugar wrack attaches itself to rocks or other large algae at or just below minus-tide mark. It is found in temperate latitudes around the world, favoring quiet waters. It is a perennial species very common in the north Pacific.

The sugar wrack is easily recognized by the broadly oval or wedge-shaped blade, short stipe, rich yellow-brown color, and thin texture. The blade may be 3-6 feet long with the width about ¼ the length. The solid stipe is cylindrical, varying in length from 1 to 3 inches. The stipe flattens out into a broad blade which is perfectly smooth except for large ripples along the margin. These are thicker than the central part of the blade. Old blades often show alternate elevations and depressions inside the ruffled margins. Just under the surface layer of cells are a number of mucilage ducts which make the blade sticky. The principal growing region of the blade is the zone between the stipe and the blade. During the nongrowing season a new blade is formed. As the new blade enlarges, the old one becomes a mere appendage at its apex and eventually breaks off.

The sugar wrack is held to the substratum by a cone-shaped mass of fibers 3-6 inches in diameter. The larger fibers branch many times and may extend 2-3 inches up the stipe. Further whorls of fibers are formed at successively higher levels as the plant grows.

NOTES ON SEA AND SHORE

Laminaria saccharina is called sugar wrack because it is sweet to the taste, owing to the presence of a sugar alcohol called mannitol. A rather wide study of the food reserves in the brown algae has been made. Those known at the present time include the polysaccharide, laminarin; the alcohol, mannitol; and fats. Mannitol forms a surface deposit of needles on the dry thallus. The mannitol content is less in the winter than in the summer and is also dependent upon depth. One investigator reported as high as 36 per cent mannitol by dry weight in plants growing in 24 feet of water.

Throughout the world there are approximately 240 genera and 1,500 species of brown algae.

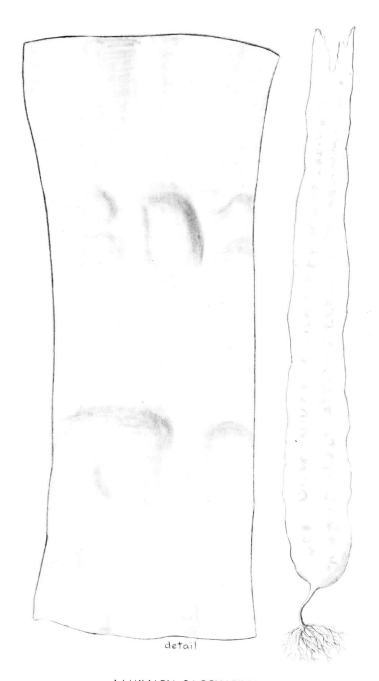

LAMINARIA SACCHARINA

Left, two-thirds natural size;
right, one-fifteenth natural size

Laminaria bullata Kjellman   Blister wrack

According to Setchell and Gardner, there are several subspecies of *Laminaria bullata,* each varying considerably in size, shape, and markings, just below the low-tide mark in quiet waters. Their range is from Alaska to Puget Sound.

The distinguishing feature of *Laminaria bullata* is the distinct row of blisters (bullae) an inch or two within the margins. The blisters are about an inch wide and an inch long. At first the short stipe (1-2 inches) is cylindrical, but it soon flattens out into a single blade which in the subspecies, *Laminaria bullata amplissima,* is from 3 to 9 feet long and from 2 to 4 feet wide, broadly wedge-shaped with the widest part above. Frequently the blade is deeply cut into a few broad segments. In the stipe, mucilage ducts are arranged in incomplete circles, while in the blade the ducts are conspicuous masses of secreting cells, making the surface of the blade very slimy. The blade is a dark olive-brown and as thick as tough rubber. The holdfast is a mass of branched fibers.

NOTES ON SEA AND SHORE

Many species of algae could be cultured on a large scale as a source of food, but so far no systematic search has been made to find the most suitable species. In the United States, Germany, Japan, and Venezuela a good deal of research has been conducted on *Chlorella,* a unicellular alga found in fresh water and soils. In 1951 experiments carried out in Maryland showed the value of *Chlorella* as a nutrient for young chickens. When 10 per cent of *Chlorella* was substituted for an equal amount of soy bean meal in a diet deficient in riboflavin, vitamin B12, and vitamin A, marked improvement in the growth of the chickens was noted. The improvement was attributed to the high riboflavin and carotene content of the *Chlorella.* In Venezuela algal soups made of phyto- and zooplankton have been fed to leprous patients. In the majority of cases there was a marked improvement in general health.

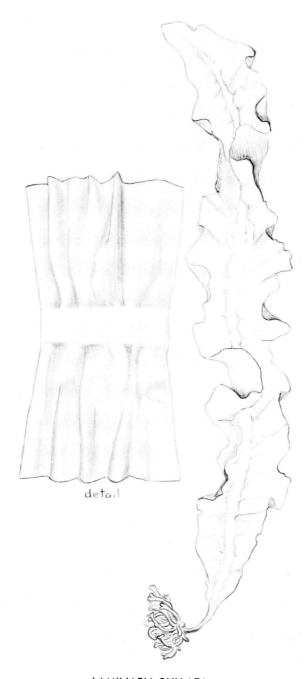

LAMINARIA BULLATA

Left, two-thirds natural size;
right, one-sixth natural size

Laminaria andersonii Eaton   Split whip wrack

As the tides and waves beat against the rocky shores, the split whip wrack is ruthlessly knocked down, but as soon as the wave recedes, it rises and spreads its split blade upon the waters, only to be knocked down by the next wave. This alga is found from Whidbey Island, Washington, to central California between the 0.5- and the -1.5-foot tide levels.

The split whip is fastened to the substratum by many branched fibers, the mass being 3-4 inches in diameter. The entire plant—stipe and blade —may be 3-4 feet long. It has a cylindrical stipe 18-24 inches long and an inch or more in diameter. The single blade is 18-24 inches long and 12-18 inches wide. The dark brown blade splits vertically into 5-10 linear segments, each segment 12-15 inches long and about 3 inches wide. Mucilage ducts are present in both the stipe and blade, making the alga very slimy. The stipe is perennial, the blade annual. When fruiting takes place late in autumn, spore cases produced in irregularly shaped patches nearly cover both surfaces of the blade. As in all *Laminaria* the growing region is the transition zone between the stipe and blade. The new blade first appears at the top of the stipe as a slight widening, separated by a marked constriction from that of the previous year. As the new blade enlarges, the old one becomes a mere appendage and eventually breaks off.

NOTES ON SEA AND SHORE

Algin is one of the important reserves in the brown algae which has come to have a wide extraction and utilization. It is an alginic acid obtained in the form of sodium alginate. The cell walls of the brown algae are made up of an inner cellulose portion called algin. Algin is used in a number of industries, particularly as a "protective" colloid helping to keep particles in suspension and emulsion. In the making of ice cream, where it is widely used, 2½ pounds of algin are used to stabilize 300 gallons of the mixture.

LAMINARIA ANDERSONII

One-third natural size

Pleurophycus gardneri Setchell and Saunders   Sea spatula

The sea spatula lives in the minus-tide zone from Alaska to southern Oregon or northern California where the currents run swiftly.

The most conspicuous feature of the sea spatula is the broad band that extends the length of the blade. The blade is 18-30 inches long and 4-15 inches wide; the midrib is 2-6 inches wide. The blade is broadest at the base, narrowing somewhat toward the upper end. Arising from the central band or midrib are broad, fully ruffled margins. The ruffles spring from regular shirrings on each side of the band. The undivided blade is smooth and elastic, more or less eroded at the upper end. In mature specimens the long stipe (10-20 inches) is cylindrical at the base, but it gradually flattens until it merges into the midrib. This alga is without mucilage ducts. The reproductive materials are in narrow patches on both sides of the midrib. As far as is known the spatula is an annual plant. The seaweed is a dark olive with the texture of medium weight rubber gloves. The holdfast is composed of several whorls of strong branched fibers.

NOTES ON SEA AND SHORE

The salt content of the oceans is generally between 33 o/oo and 37 o/oo, with 35 o/oo considered an average for all oceans. The surface salinity may be considerably less in high latitudes or in regions of high rainfall. In isolated seas such as the Red Sea, where evaporation is excessive, the salinity may reach 40 o/oo. (The salinity is given in grams per kilogram of sea water, that is, in parts per thousand for which the symbol o/oo is used.) No matter what the salinity of the water, a greater or smaller number of animals, depending upon their tolerance to salinity, adapt themselves to it.

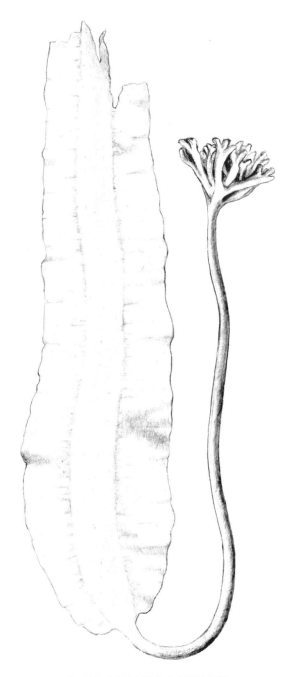

PLEUROPHYCUS GARDNERI

One-fourth natural size

Cyamathere triplicata   **Triple rib**
(Postels and Ruprecht) J. Agardh

The triple rib kelp is present in large quantities on wave-beaten shores from Alaska to Puget Sound at low-tide mark, although it is sometimes dredged from depths well below the tide mark.

Three (occasionally five) conspicuous, longitudinal ridges and grooves alternate on the upper and lower surfaces of the smooth blade, giving it a well-groomed, orderly appearance. The triple rib kelp has an exceedingly short, stout stipe, cylindrical below, flattened and slightly wider at the point where the blade begins. The blade is long and narrow, 4-6 feet in length and 6-8 inches in width, the longitudinal ridges extending the length of it. The blade sometimes splits into straplike segments at the tip because of the action of the waves and contact with rocks. It is a rich yellow-brown with the texture of heavy rubber gloves. At the base the blade is rounded, at the apex sharply pointed. Reproductive patches, called sori, are confined to one surface and to the lower part of the blade. Pronounced mucilage ducts on the blades make the seaweed sticky. The holdfast is a conical disc an inch or two in diameter.

NOTES ON SEA AND SHORE

The triple-ribbed seaweed, like the other large brown kelps, is partial to seas free of ice in the Arctic and Antarctic oceans with the north Pacific the center of its distribution. Although less variable than temperatures of the air, the temperature of the sea water plays an extremely important role in the distribution of the algae. Along the coasts the temperature variations are much greater than in the open oceans. These variations in temperature can bring about a migration of algae from one level to another in order to meet conditions favorable to growth.

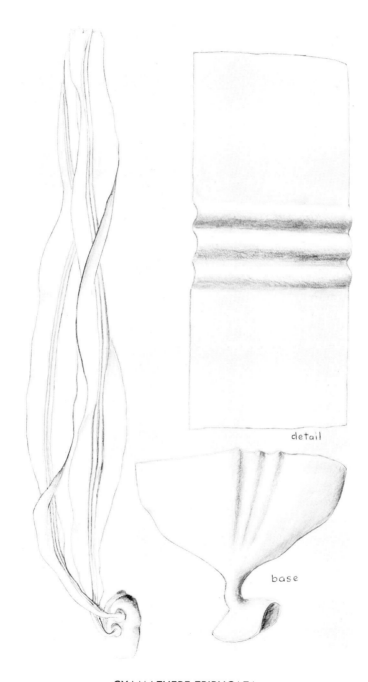

CYAMATHERE TRIPLICATA

Left, one-twelfth natural size;
right, two-thirds natural size

Costaria costata (Turner) Saunders   Seersucker

From Alaska to southern California, the seersucker kelp adorns wood, rocks, or other large algae on exposed beaches at or just below low-tide mark.

Outstanding characteristics of the seersucker kelp are the five prominent ribs which extend the length of the blade, each projecting on one side only and alternating, 3 on one side and 2 on the other. In mature specimens the portion between the ribs is puckered or shirred, but the edges of the blades are smooth. The plant is dark brown when alive but turns green when it dies. Great variety is found in the size of the seersucker kelp. In some specimens the large, flat blade may be 3 feet long and 12 inches wide. In others the length may be 8-10 times the width; in still others only 4 times the width. In the broad specimens the blade is egg-shaped. The short stipe, which is coarse in exposed areas but delicate in quiet waters, is without mucilage ducts. At the point where the stipe broadens into the blade it has many longitudinal grooves. Both the stipe and the holdfast are perennial. The growth apparently takes place in the widest part of the blade near the base. It is reported that the growth by day is double that by night, the average daily increase being about an inch. The alga fruits from midsummer until late autumn. The fruiting bodies are situated in the shirred part of the blade. The holdfast is a cone-shaped mass of wide-spreading fibers.

NOTES ON SEA AND SHORE

The seersucker kelp, like many of the large seaweeds, becomes established easily. One writer reports that, in less than nine weeks after a ship had been wrecked in the San Juan Islands, the hull was covered with full-grown seersucker plants, as well as with small algae of all descriptions.

COSTARIA COSTATA

Left, one-half natural size; right, above, one-half natural size; right, below, one-fourth natural size

Agarum fimbriatum Harvey Sea colander

The sea colander attaches itself to rocks or wharf pilings from low-tide mark to depths of 50 feet from Alaska to California, but it is particularly common in Puget Sound.

A prominent midrib about an inch wide extends like a smooth band through the entire length of the single blade. The blade, which may be 2-4 feet long and 18 or more inches wide, widens quickly from a short stipe until it becomes nearly circular or elliptical. It is thin and crinkly, perforated with holes near the edges. The perforations are due to rapid multiplication of cells, causing protrusions on one side of the blade and depressions on the other side. Another interesting feature of the sea colander is the way the blade unfolds like a scroll on either side of the base. The margins of the blade are finely toothed, giving the effect of a minute ruffle or fringe. The flattened stipe is about 3 inches long, with branched outgrowths near the base. This seaweed is yellow-brown, and has the texture of tissue paper. Fruiting bodies (sori) are formed on either side of the blade. The holdfast is a cluster of many-branched outgrowths. The sea colander is an annual plant.

NOTES ON SEA AND SHORE

The sea colander is one of the seaweeds widely used by the Japanese for food. In Osaka, kombu, a popular seaweed food, is made from the sea colander in more than fifty factories. The seaweed is gathered in open boats, then spread on the beach to dry. When all the moisture has evaporated, the seaweed is trimmed and packed into long flat bundles and shipped to the manufacturer. Kombu is cooked with meats; powdered and used in soups, sauces, or with rice; served as a vegetable; steeped as a drink; or coated with sugar and eaten as a confection. The Japanese use more than twenty species of algae as food.

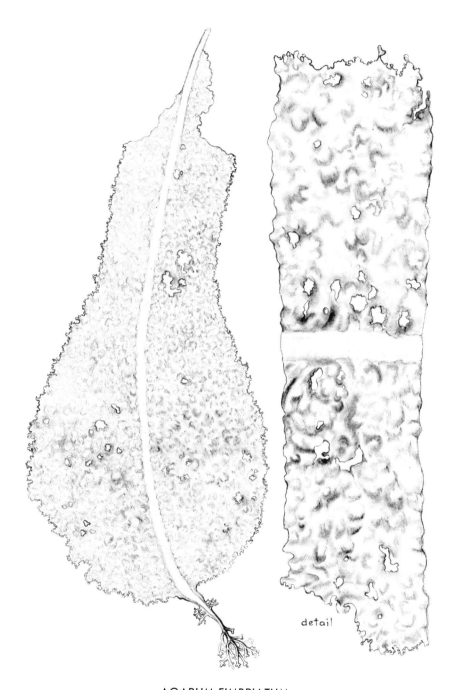

AGARUM FIMBRIATUM

Left, one-fourth natural size;
right, two-thirds natural size

Hedophyllum sessile (C. Agardh) Setchell   Sea cabbage

The sea cabbage ranges from Alaska to central California. It is especially common on rocky shores where the tides beat hard. On exposed areas it forms dense beds near low-tide mark and sometimes is the dominant species of the shore.

The sea cabbage is a mass of wide blades arranged in such dense clusters that the general effect is that of a head of cabbage. The almost circular blade is 15-18 inches long and about the same width. When the seaweed is young it has a short stipe, but this soon degenerates, and the blade seems to spring directly from the holdfast. The holdfast is composed of short, closely crowded fibers formed by the thickened basal margins of the blades. The holdfast sometimes spreads over an area of 6-8 inches. At first the blades are entire, but they soon become irregular and split through the action of the waves. The blades, either smooth or wrinkled, are always covered with large mucilage ducts. The size and markings depend upon the habitat in which the seaweed grows. The sea cabbage is a dark brown-green with the texture of heavy leather. Fruiting bodies (sori) are present on both sides of the blade near the base.

NOTES ON SEA AND SHORE

At Neah Bay, Washington, and Sooke Harbour, British Columbia, where the sea cabbage grows in profusion, the blade of the alga may be 3 feet long and 2 feet wide, folding back and forth at the base as it widens and attaches to the substratum. At these locations the rocks near low-tide mark are completely covered with sea cabbage. If one raises the blades and peers under them, one sees dozens of small seaweeds and hordes of tiny animals taking shelter under the protective covering of the sea cabbage. This protection makes life possible for many plants and animals which could not endure the direct beating of the waves.

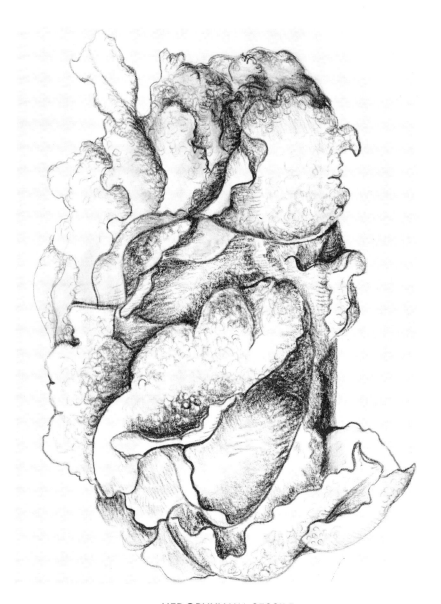

HEDOPHYLLUM SESSILE

One-third natural size

Nereocystis luetkeana (Mertens) Postels and Ruprecht

Bull or ribbon kelp

The ribbon kelp is perhaps the most conspicuous seaweed from Alaska to southern California. Although it is anchored 30-50 feet below the surface, the blades are held at the surface by air bladders. When the tide is slack, the ribbon-like blades hang down, leaving only the air bladders on the surface. When the tide is full, the ribbons float with the moving waters. This striking alga grows in large fields known as kelp beds far out from shore.

The ribbon kelp may have a stipe 30-40 feet long, the lower end solid, the upper end hollow. It is this long stipe that gives the seaweed the name bull whip. As the stipe approaches the surface, it gradually widens and terminates in an air bladder 4-6 inches long. The air bladder is filled with a couple of quarts of ordinary air and carbon monoxide which buoy up the plant. From the upper end of the bladder a number of flattened forks develop. The end of each fork continues into a long, smooth, undivided blade or ribbon 10-12 feet long and 3-5 inches wide. The blades of mature plants are shiny, dark brown, and quite thin, but in older specimens they are thick and leathery. In mature plants the stipe and blade may measure from 50 to 70 feet. Occasionally young plants 3-5 feet long may be found on the shore. These have been torn from their holdfasts and swept onto the beach. The growth of the ribbon kelp is very rapid, about an inch a day in July and August and probably more than that during the earlier stages. This remarkable growth is the result of a single season's increase, for most investigators believe that this species lives only one year.

NOTES ON SEA AND SHORE

Certain properties in the ribbon kelp are used in the preparation of pharmaceutical supplies, dairy products, poultry feeds, and glazing and finishing agents. The ribbon kelp is also an available source of potash salts. Its dry weight consists of 27-35 per cent of potassium chloride.

NEREOCYSTIS LUETKEANA

One-ninth natural size

Postelsia palmaeformis Ruprecht — Sea palm

The sea palm is a large, conspicuous alga that ranges from Vancouver Island to central California. It thrives in the upper tidal level where the surf beats hard. Occasionally large numbers of plants grow together, forming groves in crevices of rocks, inaccessible to collectors.

The plant has an olive-brown, glossy stipe, 2 feet long and 1-2 inches in diameter, tapering slightly from the base to the top. From the upper end of the stipe arise many short cylindrical branches (pedicles), each terminating in a ribbon-like blade 8-12 inches in length and 1-2 inches in width. Both the branch and the blade split longitudinally into two branches and blades of equal size. There may be as many as 100 pendent blades. Conspicuous features of the blades are the sharply toothed margins and the deep longitudinal grooves, those on one side alternating with those on the other side. The reproductive bodies are born in the grooves.

When struck by the pounding waves, the elastic stipe is bent almost horizontal but regains its upright position as soon as the water recedes. Considering the force of the waves which it resists, the sea palm has a relatively small holdfast of stout-branched fibers. The stipe is perennial, but the blades are renewed each spring.

NOTES ON SEA AND SHORE

The sea palm is reported to live only in the temperate regions of the Pacific Ocean. Like other seaweeds it is most attractive in the spring, when the blades are whole and unblemished. Late in the summer they become eroded and torn through the pounding of the surf and through the fruiting process. The graceful sea palm lends itself to decorative design and is often reproduced on the covers of books about seaweeds.

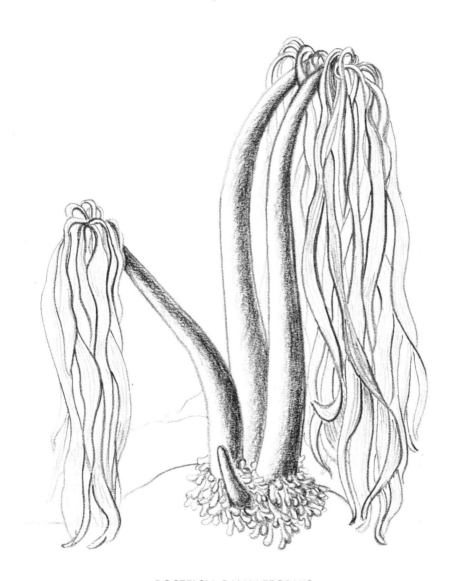

POSTELSIA PALMAEFORMIS

One-third natural size

Macrocystis integrifolia Bory　　　**Small perennial kelp**
Macrocystis pyrifera　　　　　　　**Giant perennial kelp**
(Linnaeus) C. Agardh

The perennial kelp, a world-wide seaweed, is found on the ocean beaches all up and down the Pacific Coast. It is a plant of the outer shores exposed to strong wave action. It is seen on the beaches only when it has been torn loose from its holdfast, for it lives far from shore in water 35-50 feet deep. Here it forms great beds several miles in area.

Two species of *Macrocystis*, distinguished mostly by size, are found. *Macrocystis pyrifera* may be 75 feet long in California, but the largest specimen gathered on the Washington beaches is 40 feet. The great length is brought about by the asymmetrical splitting of the terminal blade from base to apex. The cylindrical stipe, several to many feet long, is branched into a number of long, unequal parts. At regular intervals along the branched stipe are unilateral blades, which have short stipes and pear-shaped bladders. The air bladders suspend the kelp at the surface of the water. A single blade 1-2 feet long and 3-5 inches wide extends from each air bladder. The blade has an irregular, corrugated, shiny brown surface and sharply toothed margins. When the perennial kelp is washed ashore, the stipes are often matted and twisted in a most complicated manner. The holdfast is either a conical mass or a forked rhizoid from which extend densely matted branches. One writer says the holdfast is "as large as a bushel basket."

NOTES ON SEA AND SHORE

In California the perennial kelp is being used extensively in the making of algin. Since the seaweed grows in great beds, it is harvested with an underwater mowing machine carried by a motor-driven barge that holds 300 tons of seaweed. The seaweed is hoisted aboard the barge by an inclined chain conveyer. Great quantities of kelp must be cut because when fresh it contains 92 per cent water. There are about 44 pounds of algin to a ton of kelp.

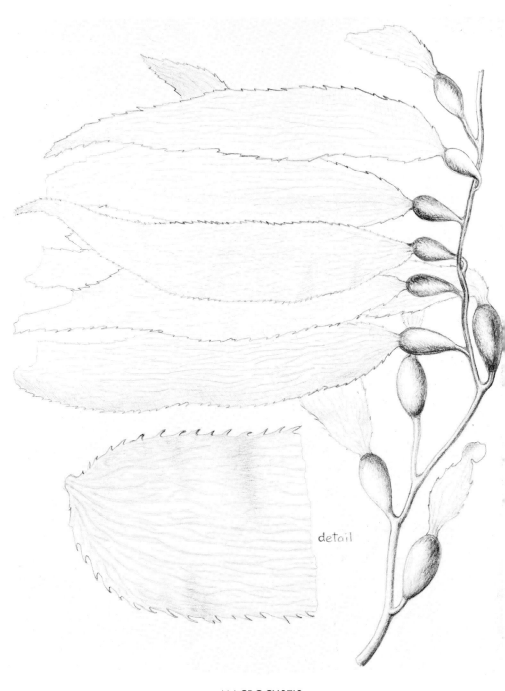

MACROCYSTIS

One-third natural size

Pterygophora californica Ruprecht   Pompon

The pompon kelp lives in extensive beds well below tide mark from Vancouver Island to Lower California. It is partial to deep crevices in the rocks where the waves beat hard.

The pompon looks like a pompon waved at football games—a long stick to which a number of streamers are attached. Although the blades and sporophylls are shed each year, the stipe may live for twenty-five years. The age is determined by the annual rings in the stipe. The stipe, about 2½ feet long and 1-2 inches wide, is woody, exceedingly stout, erect, unbranched, cylindrical at the lower end, and flattened at the upper end.

It ends in a single, tapering, narrow blade which is thickened through the entire length. Arising laterally from the upper edges on each side of the stipe are a large number (up to 40) of streamers called sporophylls. The blade and sporophylls are about 2½ feet long and 1-2 inches wide, and one must look closely to distinguish them. Both are without midribs and are perfectly smooth. The entire plant is sometimes 6 feet or more in length. Both the stipe and the streamers are covered with mucilage ducts and are dark brown, becoming black when dry. The fruiting bodies cover the surfaces of the sporophylls about the time they are shed in October. A new blade and sporophylls develop on the stipe at a higher level the next year. The holdfast is a mass of short, branched, matted fibers which hold the seaweed firm in spite of the pounding of the roughest seas.

NOTES ON SEA AND SHORE

The name kelp is the popular term used for the order of Laminariales to which belong the largest seaweeds known to man. The term also refers to the ash of seaweeds. The origin of the name is unknown, but it has been used for centuries by sailors and fishermen. It also appears in ancient sailing directions and scientific treatises. The Pacific Coast shows a great diversity of kelp. These huge plants are more abundant in arctic regions than in southern waters.

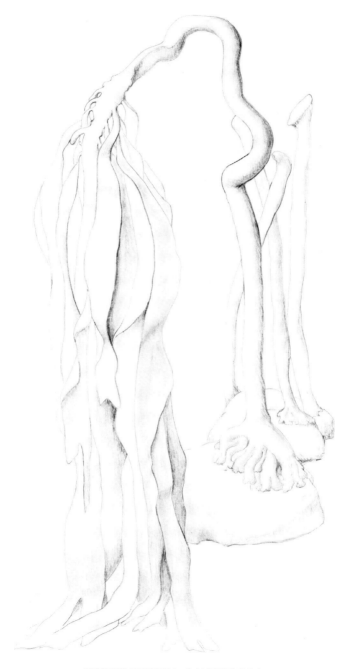

PTERYGOPHORA CALIFORNICA

One-fourth natural size

Alaria valida
Kjellman and Setchell

Honey ware or wing kelp

The honey ware, a perennial kelp often washed ashore, is widely distributed from Alaska to Puget Sound. It forms large beds in rough water 15-25 feet deep.

A single large blade with a conspicuous midrib and a number of small blades (sporophylls) are the conspicuous features of *Alaria*. The short, cylindrical stipe (2-8 inches) is perennial, the blade annual. During the summer many small blades (sporophylls) 4-5 inches long form a circle at the upper end of the stipe. Above these the stipe flattens out and terminates in a single blade 3-12 feet long and 8-15 inches wide. Extending the length of the blade is a thick midrib ¼-¾ inches wide. The surface of the blade is smooth with slightly waving margins. The blade widens gradually to the middle, then narrows to a rather sharp point. The reproductive bodies, called sori, cover the small winged blades (sporophylls) except for a narrow margin around the outer edge. After the sori mature the sporophylls drop off, leaving scars on the stipe. New sporophylls grow out above these the next year. The wing kelp is a clear, almost transparent yellow-brown with the texture of medium-weight rubber. The holdfast is a circular, wide-spreading mass (3-4 inches) made up of many short, solid fibers. In Alaska some species of *Alaria* are 30 feet long.

NOTES ON SEA AND SHORE

The smaller, more delicate algae usually reproduce during the summer and early autumn, while the larger ones such as *Alaria* reproduce during the winter months. The honey ware or wing kelp is reported to have a remarkable growth during the winter months, the daily average winter growth being ¾-1 inch, with the rate of growth during the day twice that during the night. This seaweed was formerly eaten in Scotland, where it is called badderlocks, and in Ireland, where it is called mirlins.

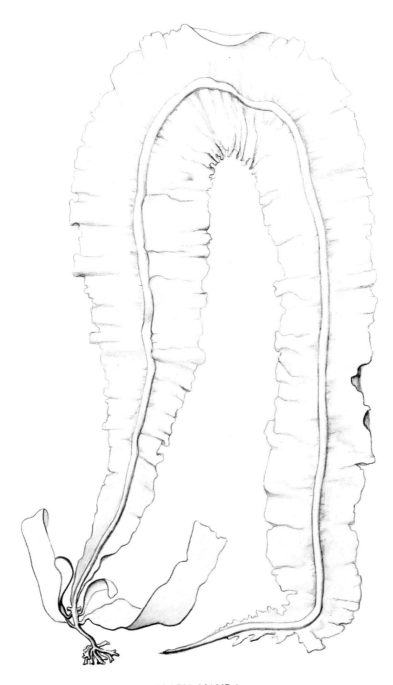

ALARIA VALIDA

One-fourth natural size

Egregia menziesii (Turner) Areschoug   Feather boa

The feather boa is a perennial kelp found on the surf-swept coasts from Vancouver to central California.

From the short (2-3 inches), cylindrical stipe, which soon becomes flat and irregularly branched, arise 6-25 main branches, each branch continuing in a long, narrow, stemlike growth. At irregular intervals smaller branches arise from the main branches. The outer edges of all the branches are densely fringed with short (2-3 inches) blades of various shapes and sizes. Some of the short blades develop into spindle-shaped air bladders which buoy the seaweed on the surface, some are flat and oval, and some are wedge-shaped. The last bear the reproductive organs. The surface of both the stipe and the blades is covered with small, blunt tubercles. Mature plants may be 15 or more feet long with branches 3-4 feet long. From these arise the short blades, which are approximately 6-8 inches wide including the dense outgrowths extending from the edges. The feather boa is a dark chocolate-brown, almost black. It has a compact, cone-shaped, profusely branched holdfast 6-7 inches in diameter growing around the base of the stipe.

NOTES ON SEA AND SHORE

In some parts of the West Coast, *Egregia* forms large beds at low-tide mark, the branches being twisted and matted in a complicated manner. Like the other large brown algae, this species is used for fertilizer. For many centuries seaweeds have been used to increase crop yields in countries with extensive seacoasts. The fertilizing properties of seaweed are about the same as those of barnyard manure. Tests have shown that seaweeds contain a greater proportion of potassium salts, about the same proportion of nitrogen, but a smaller amount of phosphoric acid than does manure.

EGREGIA MENZIESII

One-half natural size

Fucus furcatus C. Agardh   Rockweed or popping wrack

The rockweed lying high on the beach constitutes almost half of the seaweeds between Alaska and central Calilfornia. However, it is a genus difficult to describe because of its many variations.

The rockweed is characterized by flattened, forked branches lying on one plane, more or less prominent midribs extending through the entire length of the alga, swellings at the ends of the branches, the ability to remain for a long time out of water without drying out, and resistance to changes in temperature. The cylindrical stipe, 1-2 inches long, merges into a flattened blade which forks several times at intervals of an inch or more. The pattern of branching is quite uniform. The green-brown to almost black rockweed averages 10-12 inches in length. The branches near the stipe are 1-2 inches in width, and those toward the upper end are longer and narrower. The branches are rounded at the ends, are more or less sticky, and have swollen bladders at the ends. The swelling is caused by the presence of mucilage.

The male and female reproductive bodies are also carried within the bladders. When the bladders are compressed, water squirts from tiny holes, producing a popping sound. The holdfast of the rockweed is a small disc from which several stipes arise.

NOTES ON SEA AND SHORE

Fucus (rockweed) has 64 sperms in the male organ and 8 eggs in the female organ, the egg being as much as 30,000 times larger than the sperm. During ebb tide the small, motile sperms, and the motionless eggs are discharged into the water. While the eggs float, swimming sperms congregate near them, attach themselves, and spin around at the point of attachment. One sperm fertilizes the egg, and the others disappear. The fertilized egg immediately begins to develop. *Fucus* produces eggs and sperms throughout the year. Of course this process cannot be seen with the naked eye.

FUCUS FURCATUS

One-half natural size

Cystophyllum geminatum
(C. Agardh) J. Agardh

Bladder leaf

The abundant bladder leaf often forms a conspicuous belt just below low-tide mark in sheltered places from Bering Sea to Puget Sound.

This seaweed is 3-4 feet long with soft flexible branches 18-24 inches long and ⅛ inch wide, extending at fairly regular intervals from a prominent main axis. From the primary branches hang long, narrow, lax secondary branches ending in short, toothed leaflets borne on slender stems. Tiny 1/16-⅛ inch air bladders like miniature balls are suspended from very short stems on the secondary branches. These help to suspend the seaweed at the surface of the water. The secondary branches, which subdivide several times, are about 7-8 inches long. The lower branches are short, flat, and leaflike; the upper branches are long and almost threadlike. The bladder leaf is a dark olive-brown, often showing a lovely iridescence when under water. The holdfast is a solid, fibrous, conical disc.

In *Cystophyllum* the single-celled female sex organ, which contains one or more eggs (oogonium), remains attached for several days to the inside of the conceptacle by a long gelatinous stalk before being pushed into the water where it is fertilized. It is believed this genus of algae is still in the process of vigorous evolution. Most of the species of *Cystophyllum* are found in Japan.

NOTES ON SEA AND SHORE

Rhythmic periodicity of sex organs has been observed in the bladder leaf and related genera of seaweeds. In many regions 60-70 per cent of the sexual cells are released in a single hour before daybreak. This occurs in some regions once every two weeks, in other regions once a month.

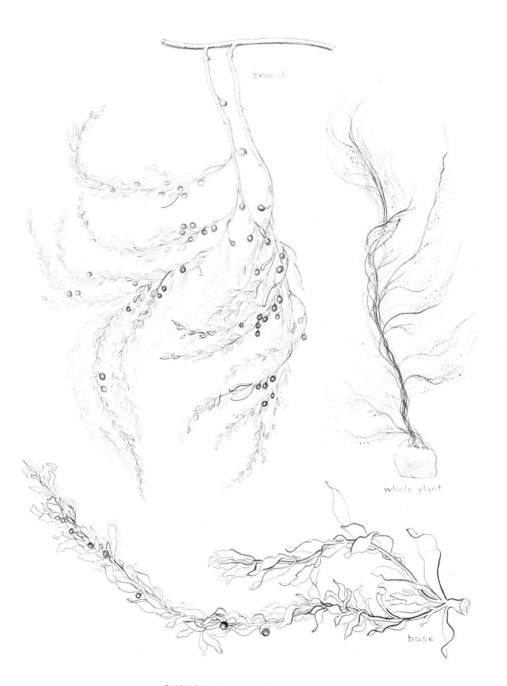

CYSTOPHYLLUM GEMINATUM

Whole plant, one-twentieth natural size;
branch and base, two-thirds natural size

Cystoseira osmundacea
(Menzies) C. Agardh

Woody chain bladder

The woody chain bladder is common between the mean low-tide and the -1.5-foot tide levels from Washington to Lower California; however, it is often dredged from depths of 20-30 feet. This seaweed may reach a length of 18-30 feet. The lower part is blackish brown and the upper part light tan.

The species has a strong, conical holdfast, an inch or more in diameter, from which arises a woody stipe. The short, angular stipe gives off several shoots from which arise the primary branches, several feet long. Secondary branches arise from the primary shoots. Both are flattened at the lower end, and the upper portions of both are cylindrical. On some of the smallest branches are leaflets 1 inch long and ¼ inch wide. A chain of 5-12 tiny bladder-like sacs is attached a short distance back from the sharp tip of the leaflet. The sex products are produced within cavities at the tips of the ultimate branches. The female organ produces a single egg wih seven nuclei; the male organ produces and liberates (64?) cells. The stipe and the primary branches are perennial, but the smaller branches and the air bladders disintegrate after the plant fruits.

NOTES ON SEA AND SHORE

Cystoseira belongs to the famous and baffling Sargassum family. Sargassum is one of the most widely known seaweeds because of the great beds, known as the Sargasso Sea, lying in the middle Atlantic. The sargassum is a brown alga which grows along the coast of Florida and the West Indies. Many of the plants are torn away by storms and carried northward by hurricanes until they reach the center of the Atlantic Ocean. Because there are no winds and no strong flow of water there, the seaweed is caught and held indefinitely. It takes several years for the seaweed to be carried into the center of the area.

CYSTOSEIRA OSMUNDACEA

Detail, natural size; others, two-thirds natural size

Porphyra naiadum Anderson   **Red fringe**

On sandy beaches and in quiet waters from Vancouver Island to southern California, the red fringe hangs from the edges of eel grass and other seed-bearing marine plants like a delicate fringe or ruching. The fringe is purplish red to deep purple, while the blades of the host eel grass are dark green. The 3-5 blades of each alga are about ½-1 inch long and ½ inch wide at the upper end, narrowing at the base until they become stipelike.

Because the red fringe has only one layer of cells, it is as fragile as the thinnest tissue paper. The cells are embedded in a colorless gelatinous sac, with the color pigment located in a large, specialized portion of the protoplasm. The cells near the base grow downward and thicken to form the holdfast. The microscopic holdfast is made up of numerous interwoven threads that are outgrowths of the plant body. These are capable of expansion and may produce new branches. The red fringe is an annual plant, appearing early in the spring and disappearing late in the autumn.

NOTES ON SEA AND SHORE

The genus *Porphyra* has a wide distribution in both the northern and southern hemispheres, but it is not a highly developed genus of red algae. In fact, its exact classification is a matter of considerable discussion. It is an annual plant with a variable seasonal growth depending upon the amount of water available and the intensity of light and shade. It is particularly partial to the sun. The blades of *Porphyra* are gelatinous, thin, elastic, and satiny, and their sheen is retained when the plant is dry. Some of the larger species with elaborately ruffled margins are very beautiful when in the water, glistening and shimmering like red satin.

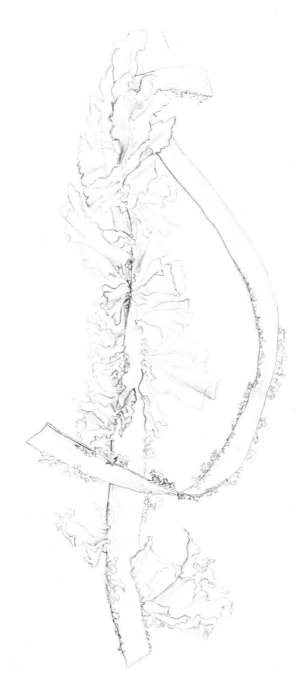

PORPHYRA NAIADUM on EEL GRASS

Natural size

Porphyra perforata J. Agardh   Red laver

Red laver is common everywhere between the 3.5- and the 2-foot tide levels from Alaska to southern California.

When floating on the water, the red laver is beautiful, the single flattened blade with the ruffled margins swaying on the restless waters. When out of the water, it is dull and slimy. When the plant is young, it is a bright red, the blade narrow and tapering with deeply ruffled margins; when older, it becomes a grayish purple, the blade broad, deeply slashed and torn. Both the shape and the size vary greatly, depending on the climatic conditions, but frequently the blade is 18-20 inches in length and width. The single blade, attached by a tiny disc, is without markings and is extremely fragile, having but one layer of cells. However, it becomes practically two layers when the reproductive bodies begin to form along the margins of the blades. This thickening takes place through the swelling of the jelly surrounding the reproductive bodies at the time of ripening. Thousands of spores are shed when they become mature. If they lodge on suitable surfaces and the conditions are favorable, germination takes place at once, and a new plant appears the next year.

NOTES ON SEA AND SHORE

In *Marine Products of Commerce,* Donald Tressler states that of all seaweeds that are dried and used for foods (omitting seaweed extractives) the laver industry is the largest. On the Pacific Coast of the United States, Chinese-Americans process many thousands of pounds of red laver a year, exporting some of it to China. In Puget Sound the Indians still use porphyra as flavoring for soups and meats. In Japan the Japanese farmer sets out bundles of bamboo poles in shallow water or sheltered shores. On these poles the spores of the laver are caught and supported while they develop. The bundles are set out in the autumn, and in a few months the laver is ready for harvest.

PORPHYRA PERFORATA

Two-thirds natural size

Porphyra lanceolata   Red jabot laver
(Setchell and Hus) G. M. Smith

The red jabot, ranging from Puget Sound to central California, is a beautiful seaweed.

As the thin, ruffled, spirally twisted red blade undulates with the quietly moving waters on a sandy beach, it looks like a frilly jabot such as women used to wear on their blouses. The red jabot may be 3-9 feet long and 6-12 inches wide, the male plant being narrower than the female. In the male plant the blade narrows toward the upper end; in the female the broad tip is usually deeply cut and torn. There is only one layer of cells in the red jabot. This is gelatinous and elastic, with a satiny sheen. The sheen and the red color are retained when the plant is dry. Most species of *Porphyra* are annual plants partial to the sun. The sessile blade has a tiny disc-shaped holdfast. Both the male and the female products are in pockets on the margin of the blade. The red jabot has a beautiful iridescence—shades of green, blue, and purple—as the sun plays upon it.

NOTES ON SEA AND SHORE

No one knows how the earth acquired its oceans. But it is believed that the earth, freshly torn from the parent sun, was a ball of whirling gases, intensely hot. Gradually the gases commenced to cool and to liquefy. In time the materials of the earth were sorted out in a definite pattern—the heaviest in the center, the less heavy surrounding this, and the least heavy at the outer rim. After millions of years the outer shell of the earth was enveloped in a heavy layer of clouds, which contained much of the water of the planet. Eventually the surface of the earth cooled enough for the clouds to produce rain. Rain fell day and night, century after century. It poured into the uneven basins of the earth's surface and fell upon the continental masses until at last the great oceans were made.

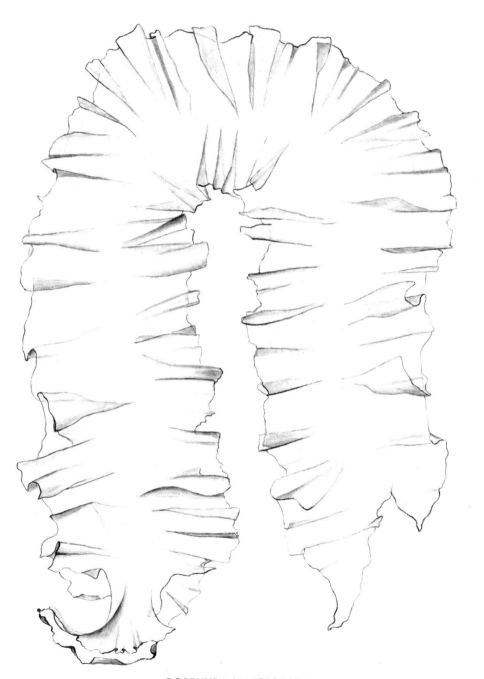

PORPHYRA LANCEOLATA

One-third natural size

Farlowia mollis   Farlow seaweed
(Harvey and Bailey) Farlow and Setchell

The Farlow seaweed grows abundantly from Vancouver Island to central California between the 0.5- and the -1.5-foot tide levels.

As this seaweed is rocked by the waters, the small, dense clusters of bright red to very dark red branches are attractive. The soft, loosely arranged branches are occasionally 6-8 inches long, although those 3-5 inches are more common. The major branches, roughly 1/16-⅛ inch wide, rebranch to the second or third degree. Usually they become progressively shorter toward the apex, although at times the wider branches are at the tips. The branches are irregularly arranged, either alternate or lying so close together they appear to be opposite. Hairlike branchlets are arranged in two rows along the sides of the branches. The branches are without midribs or veins and are flattened and slimy. The arrangement of the branches is extremely haphazard. At times so many branches arise from the disc-shaped holdfast that the plant takes on the shape of a rosette. The holdfast is often deeply buried in the sand.

NOTES ON SEA AND SHORE

The red algae show no close affinities with other classes of algae; thus they must have broken off from other forms at a very early stage. We have little knowledge of their fossil history, and what little is known sheds almost no light on their evolution. It is evident that the calcified red algae are very old; they appear continuously from the Cretaceous period and have played an important role in the formation of the limestone rocks. Two distinct layers are generally recognized in the cell walls of the red algae: the inner, firmer layer consisting of cellulose, and the outer layer composed of jelly-like substances, related to the pectins of the higher plants.

FARLOWIA MOLLIS

Natural size

Constantinea simplex Setchell Cup and saucer

From Washington to California one finds the cup and saucer seaweed a short distance below tide mark. After storms it is often cast ashore in large quantities. This seaweed is restricted to the north Pacific Ocean and the Bering Sea.

This is an unmistakable species, for it looks like a bright red cup and saucer standing on a pedestal. The effect is produced by the thick, cylindrical stipe which passes through a flat, circular blade. The stipe forms the cup, and the blade the saucer. Sometimes several saucers, each up to 4 inches in diameter, stand above one another on a stipe 3-5 inches long and $1/4$-$1/3$ inch in diameter. At first the blades are perfectly round, but later they are torn into wedge-shaped segments. They ultimately wear away, leaving scars on the stipe. The interval between the blades or scars of old blades varies with the size of the plant, but it is usually $1/4$-$1/2$ inch. Growth is by a continuation of the stipe passing through the topmost saucer. The stipes, either solitary or in clusters, arise from a disclike base. Another species of Constantinea is much smaller, with a stipe $1/8$ inch in diameter and a blade about an inch in diameter.

NOTES ON SEA AND SHORE

In proportion to the tremendous quantities of food material latent in the sea, its present use by man is infinitesimal and indirect. Its chief use is as a source of nutriments for the thousands of species of animals which live in the sea. However, by a circuitous route these nutriments reach you and me. This road is represented as passing from algae to protozoans and copepods; thence to crustaceans, such as crabs and shrimps; on to fish, such as salmon and halibut; and eventually to man. At the present time and probably for centuries to come, the ocean will be an unconquered wilderness, but further knowledge of this great frontier would be a tremendous benefit to man.

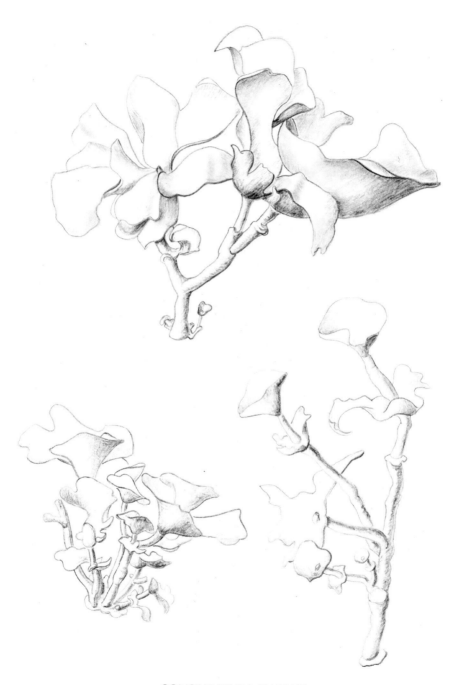

CONSTANTINEA SIMPLEX

Natural size

Endocladia muricata **Nail brush**
(Postels and Ruprecht) J. Agardh

The nail brush seaweed grows from Alaska to southern California where the surf beats strongly. It is extremely abundant, often forming an algal zone at the uppermost reach of the seaweeds. In fact, it grows so high and can remain out of water so long that it can survive when it is only sprayed by the surf.

This seaweed is a bushy, dark red or blackish-brown form with stiff, short branches. The dense branches are 1-2 inches long and 1/16 inch in diameter. Covering the surface of the cylindrical branches are harsh bushlike spines, 1/32-1/16 inch long.

The male and female organs are borne on separate plants. The male cells are in small, irregularly shaped patches on the thallus, while the female cells develop deep within the branch. The envelope in which the fertilized cell rests is urn-shaped and stands 1 mm. tall. During the fruiting season the envelope is easily seen with the naked eye, since it is much lighter in color than the branch bearing it. The holdfast is a small, strong disc from which a number of thalli arise.

NOTES ON SEA AND SHORE

Sexual reproduction in red algae is by the passive transportation of a male cell to a hairlike extension of a one-celled female sex organ. The male cells drift about in the water until the currents bring them into contact with the hairlike extension of the female cells, which retain the egg for some time. The fertilized egg, either by direct division or by the production of filamentous outgrowths, produces a number of spore sacs (sporangia), each holding a single spore.

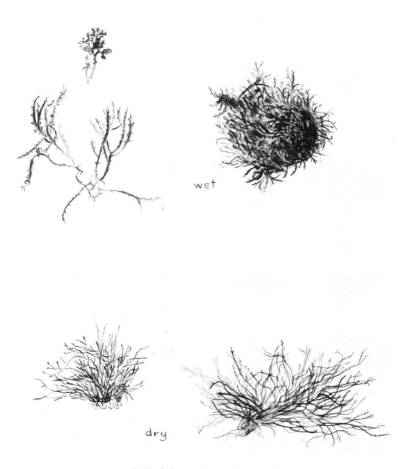

ENDOCLADIA MURICATA

Natural size

Lithothamnium Philippi   Red rock crust

The red rock crust is found quite generally up and down the Pacific Coast, where it appears as a stony encrustation on rocks or shells between the 0.5- and the -1.5-foot tide levels. Occasionally it grows in free, somewhat globose tufts.

Although this seaweed has none of the characteristics we are accustomed to think of as belonging to a plant, it is one of the easiest genera of seaweeds to recognize. However, to distinguish species is extremely difficult. The rock crust resembles a thick blotch of red paint spilled on a rock. It usually has a roughly circular shape, 1 to 4 inches in diameter, either smooth, covered with small knoblike outgrowths, or with raised plates extending across the surface. The outgrowths may be 1/16-1/8 inch high. Sometimes the plants overlap one another to form patches of indefinite extent, while others are attached only by the middle portion of the lower face. The color varies from whitish pink to deep purple. These seaweeds are perennial, sometimes living many years.

NOTES ON SEA AND SHORE

Algae, like all living things, must have food—carbohydrates, fats, proteins, and minerals. Yet perhaps the most important job of the seaweeds is to make food for animals. In the sea, as on the land, the plants are the real producers. They are the organisms capable of making complex organic substances from the simple organic compounds dissolved in water. In doing this the algae utilize complex and varying methods, depending upon the genus of seaweed and the season of the year. If it were not for the marine plants, development of marine animal life would be almost impossible.

LITHOTHAMNIUM

Natural size

Corallina gracilis var. *densa* Collins   **Graceful coral**
Corallina chilensis Decaisne   **Tide pool coral**

Two species of coral algae are especially common on rocks between the 0.5- and the -1.5-foot tide levels from Vancouver Island to southern California. *Corallina gracilis* likes rocks exposed to strong surf, while *Corallina chilensis* prefers quiet water and tide pools.

The graceful corals are calcified—hard and brittle—through the deposit of lime salts within the segments of the thallus. The portion between the segments is not calcified, however, so that the species have erect shoots with joined flexible branches. The branches of *Corallina gracilis* are 1½ to 3 inches tall. They are divided into many segments 1/32-1/16 inch in diameter. The axes of the shoots and branches are covered on both edges with short, uniform, lateral outgrowths. *Corallina chilensis* is 2-5 inches tall, freely branched several times, with segments slightly broader than those of *Corallina gracilis*. In both species the branches are on opposite sides of the axis, and on each segment of the branch there are lateral outgrowths. The segments are short, narrow, and compressed. Most of the branches lie on one plane and are close together. This genus of seaweed has a crustlike base of indefinite extent. Several shoots arise from the same crust. The graceful corals are purple-red to deep purple. When exposed to strong light or at death they bleach to white or pink. The sex organs are borne either on the same or on different thalli.

NOTES ON SEA AND SHORE

The calcified algae first appeared in the Cambrian geological period over a hundred million years ago. They have played an important part in the formation of limestone. Some authorities believe that the coralline algae have done more toward the building up of coral reefs than have the corals (animals). Perhaps the name algal reef should supplant that of coral reef.

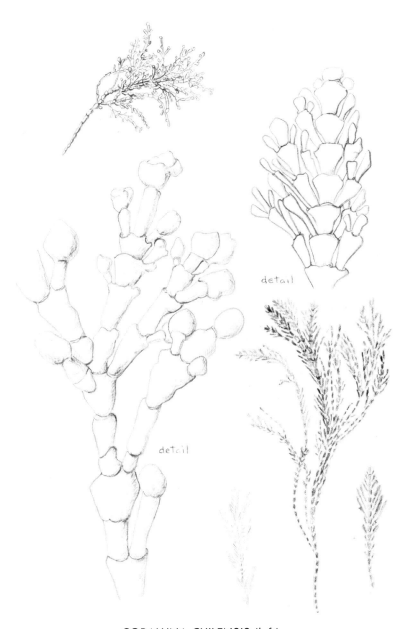

CORALLINA CHILENSIS (left)

Above, natural size; below, eight times natural size

CORALLINA GRACILIS (right)

Above, eight times natural size; below, left to right, one, two, and three times natural size

Bossea Manza **Leaf coral**

Although the leaf coral has a wider distribution in California than farther north, it is found on a number of beaches in Puget Sound and along the Washington coast. This unusual alga, which varies from 1-6 inches in height, attaches itself to rocks or shells between the mean low-tide and the -1.5-foot tide levels.

The characteristics of the genus are calcification, the many erect, jointed shoots, the forked branches, the thick, broadly rounded branchlets, and brittleness. However, it is difficult to distinguish species within the genus. The brittle branches may be on opposite edges of the shoots like a feather, they may be forked, or the same plant may have a combination of the two systems of branching. The segments of the branches are so distinct that they resemble tiny beads strung on a wire; those on the upper part of the branch are compressed, those on the lower part cylindrical. The densely crowded branches may be ½ inch long and 1/16 inch wide. Successive branches bear branchlets with prominent midribs. The branchlets are flat, densely crowded along the branches, approximately 1/16 inch long, and slightly wider than long, with the upper ends rounded. Although the segments are heavily calcified, the joints are not calcified, allowing for flexibility. This alga varies in color from pinkish white to deep purple, the variation due to bleaching of exposed plants. The leaf coral has a crustlike, lobed base from which the shoots arise.

NOTES ON SEA AND SHORE

In the coralline algae, the calcification extends to all parts of the plant except the reproductive organs and the very young growing tissue. The envelope of lime is thicker in well-illuminated forms than in those found in shaded areas. In addition, the plants usually contain a greater amount of lime in the winter than in the summer. The development of calcareous structures in plants is usually brought about when the carbon dioxide is removed from solution by photosynthetic activity.

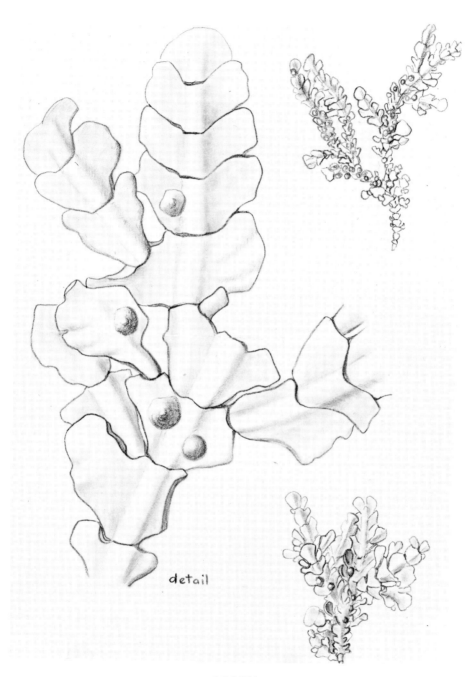

BOSSEA

Detail, twelve times natural size; other views, natural size

Calliarthron Manza
Bead coral

Many more coralline algae live in the waters off California than in Washington waters. Nevertheless, the bead coral is found on several Washington beaches, where it adheres to rocks or mollusc shells between the 0.5- and the -1.5-foot tide levels. The bead coral, like other coralline algae, has a calcified crustlike base of indefinite extent from which arise many erect, branched, jointed, flexible shoots.

The most noticeable feature of this seaweed is the resemblance to beads strung on a wire. The segments or beads are so brittle they break at the slightest touch. The erect shoots are 2-5 inches in height (possibly taller) with widely spaced forks or branches. The branches are narrow, naked, and without branchlets. The branching is loosely alternate, continuing to the third or fourth degree. The lower branches are 1/16-1/8 inch wide, the upper ones slightly wider. In general, the branches are all on the same plane. The segments of the lower parts of the shoots are cylindrical; those above are somewhat flattened and have conspicuous wings, often with deep indentations on the edges. As in the other coralline algae, the segments of the bead coral are strongly calcified while the joints between them are not calcified; this allows for flexibility. When the seaweed is alive, it is a reddish purple, but when dead or dry, it is chalky white.

NOTES ON SEA AND SHORE

Calcified algae usually live in water 150 or more feet deep. This indicates that they can survive in regions where other algae perish. Up to the end of the eighteenth century, vermifuges in the Western Hemisphere were made from several red, lime-encrusted algae. In 1775 they were replaced by a single red alga called Corsican moss *(Alsidium helminthochorton)* by a doctor working in the Mediterranean. The Chinese also used algae as vermifuges, but instead of using one kind they made a concoction of one green, one brown, and seven red algae. They believed that if one species was effective several should be proportionally more so.

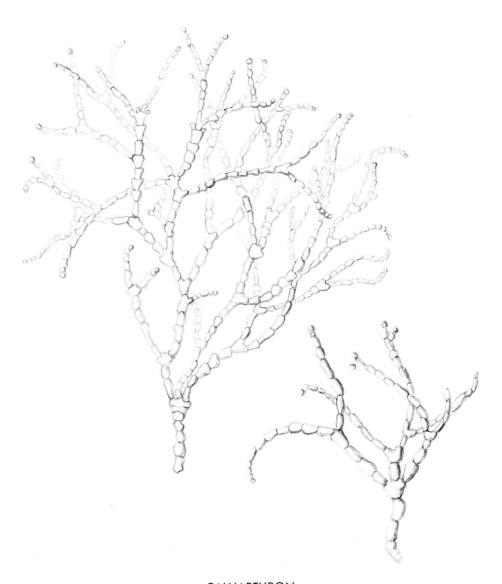

CALLIARTHRON

Left, natural size; right, twice natural size

Grateloupia pinnata   Pointed lynx
(Postels and Ruprecht) Setchell

The pointed lynx is widely distributed from Puget Sound to southern California and grows profusely on rocks in the littoral zone or at the -1.5-foot tide level. It is also found in tide pools, in polluted areas, or where fresh water enters the sea.

The most noticeable characteristics of this seaweed are the feather-like leaflets (proliferations) that extend from the edges of the flat blades, 5-7 inches long and ½-1 inch wide. The closely packed leaflets are roughly ½-1 inch long and 1/16 inch wide, all lying on the same plane. Smaller proliferations may extend from the proliferous branches. The base of the blade is narrow, then broadens out and again narrows toward the apex. The upper part of the blade often has holes and linear cuts in it, the result of weathering or of the dropping off of the sori (small patches of male cells). The blades are smooth and have a soft, gelatinous texture. With a color range from bright red to olive-purple, this seaweed is striking in appearance. The holdfast is a disc-shaped organ, with several shoots growing from it. The male and female organs are on different plants. The male products are in small, whitish patches on the surface of the blade; the eggs are borne in special bushy filaments on the outer face of the medulla (the central tissue of the thallus).

NOTES ON SEA AND SHORE

For purposes of plant and animal study the ocean is divided into three zones: the littoral (between the tides), the shallow water, and the deep water. The littoral zone is of the greatest imporance for plant study because here seaweeds carry on photosynthesis at the maximum rate. It has been found that most photosynthesis occurs in both red and green algae at depths between 15 and 30 feet. On cloudy days the maximum photosynthesis may occur at the surface. Red algae can use light to depths of 60-70 feet, but the brown algae cannot make carbohydrates at depths of more than 15 feet.

GRATELOUPIA PINNATA

Three-fourths natural size

Prionitis lyallii Harvey   Lyall's seaweed

Lyall's seaweed is abundant on every exposed beach between the mean low-tide and the -1.5-foot tide levels on rocks exposed to heavy surf from British Columbia to central California.

The surface of the flattened branches is smooth, soft, and gelatinous. Arising from the holdfast are one or more shoots which begin to flatten into the main stem when about ½ inch long. From this arise many irregularly arranged flattened branches, all of approximately the same breadth. Although exceedingly variable, the seaweed usually measures from 4 to 12 inches in height with the primary branches ½-1 inch in width. The branching extends to the second or third order. From the edges of some of the branches are numerous flattened outgrowths (proliferations). These proliferations, all on one plane, are narrow and tapering and vary from 3-6 inches in length and from 1/16-⅛ inch in width. Along the edges of the proliferations there may be still smaller outgrowths ¼-½ inch long. The holdfast is a tiny, disc-shaped organ, the discs of the different plants crowded closely together. The male and female organs are borne on separate plants. The sperm forms whitish patches that nearly cover the branches; the female organs are borne on special bushy, branched filaments.

NOTES ON SEA AND SHORE

The rocks in the intertidal zone are the most favored habitat of seaweeds; often a dozen or more species and innumerable specimens are crowded together in a complicated mass. For this reason the seaweeds living in this milieu have many problems to solve. They must juggle desperately for positions which offer optimum water and oxygen; they must adjust themselves to being left high and dry when the tide goes out, to being exposed to the hot sun during low tide, to the decreasing light of the deeper water when the tides are high, and to the grinding of the wave-washed rocks.

PRIONITIS LYALLII

Natural size

Callophyllis edentata Kylin   Red sea fan

The red sea fan ranges from Whidbey Island, Washington, to southern Oregon. It may be found attached to rocks below mean low-tide levels, dredged from water 30 feet deep, or cast ashore.

The uniform, clear, dark red blade is fan-shaped, an average-sized plant being 4-8 inches tall with the maximum width of the fan 5-6 inches. For a distance of about an inch, the blade is narrow and undivided; then suddenly it divides 4-6 times. It is this rather regular division that gives the alga its fanlike shape. The major divisions redivide again in the same manner. In general, the successive divisions are the same width, with the ends of the blades bluntly rounded. The distance between the successive divisions is about an inch. The blade is on one plane, smooth without veins or midrib, paper thin, and almost transparent. As a general rule the margins of the blades are smooth, but occasionally they are curled. The blade directly arises from a minute, disclike holdfast. In another species, *Callophyllis flabellulata*, the tips of the branches are divided into many narrow forks. The fertilized cells of both species appear as tiny dots scattered over the entire blade; some of these are large enough to be seen with the naked eye.

NOTES ON SEA AND SHORE

Man knows little of the kelps that live more than 50 or 60 feet below the surface of the water, but undoubtedly vast forests of dark kelps exist in the deep reaches of the sea. For example the great beds of ribbon kelp *(Nereocystis luetkeana)* disappear with the coming of winter, but nature has provided for its reappearance the following spring. In the middle of the summer reproductive cells appear on the ribbons. Just before the parent plant dies, the mature cells drop off. If they find suitable places of attachment and germinate, they spend the winter far down in the water out of the reach of the waves. By the first of March a tiny plant appears. As it grows it reaches higher and higher toward the surface until by the first of June it is 50 feet long. The deep waters have nurtured and cradled it until it could take its place among the swaying, golden-brown masses of ribbon kelp.

CALLOPHYLLIS EDENTATA

Natural size

Schizymenia pacifica Kylin   Sea rose

The sea rose, widely distributed from Alaska to southern California, grows between the mean low-tide and the -1.5-foot tide level on rocks exposed to the surf, or it is dredged from a depth of 30 feet.

The most noticeable features of the sea rose are the rich brownish-red color and the smooth, sticky, flat, relaxed blade which arises from an extremely short (⅛ inch), cylindrical stipe and tiny holdfast. Almost at once the stipe broadens into a full, membraneous blade which looks like the torn petals of a rose. When several blades arise from the same holdfast, they form a more or less compact cluster. The blades may be 9-15 inches tall, but the shape is variable, sometimes broadly rounded or broadly pointed, depending upon the environment. Plants exposed to the waves are usually longer than broad; those in sheltered places tend to be broader than long. In many specimens the margins are frayed or deeply cut from the beating of the waves. When partially dried the surface becomes finely granulated. The male and female organs are borne on separate thalli, with the male cells arranged in patches along the margins of the blades.

NOTES ON SEA AND SHORE

It is impossible to imagine a medium richer or more varied in basic elements than the sea. In fact, it is probable that the sea contains in solution at least traces of every element. In *The Oceans* (Sverdrup, Johnson, and Fleming), 49 elements are listed, and the authors state that further studies will undoubtedly show the presence of many more. Other investigators have tabulated 39 elements, 18 invariable and 21 variable ones. All of the atmospheric gases are also found in solution in sea water. In addition to nitrogen and oxygen, the most abundant gases, large quantities of carbon dioxide are present. Other factors that may modify the concentration of sea water are the deposits from rivers, freezing and melting of sea ice, and biological activity. Amazingly enough, the marine animals and plants put to use almost every element the sea offers.

SCHIZYMENIA PACIFICA

Natural size

Agardhiella coulteri (Harvey and Bailey) Setchell

Coulter's seaweed

Coulter's seaweed is widely distributed at the 1- to -1.5-foot tide levels in tide pools, washed onto the shore, or dredged from depths of 20 feet. The range is from British Columbia to southern California.

This seaweed is a conspicuous bright pink to dark red. Its outstanding feature is the neat, smooth surface of the many fleshy branches. The branches never mat or entwine, and dirt, debris, or other algae do not cling to them. The shoots become 12-15 or more inches long. When a number of shoots arise from the same holdfast, they form dense clusters. Usually the fleshy branches 1/16-⅛ inch in diameter are closely and irregularly arranged on all sides of the cylindrical axis, but they may lie on one plane only. The branches toward the base of the plant may be 6 inches long, but they become progressively shorter toward the apex until those at the tip are less than ¼ inch long. Each branch ends in a sharp point. Occasionally a few short secondary branches arise from the main branches, but this is the exception rather than the rule. The sex organs are on different thalli, the fruiting bodies appearing as nodules or localized swellings along the branches. These are often quite conspicuous. The holdfast is a disc-shaped organ from which one to several fleshy shoots grow. Old holdfasts may have prostrate cylindrical outgrowths.

NOTES ON SEA AND SHORE

Considerable investigation of the food properties in algae has been made, but more study remains to be done on this important subject. Algae as a class contain huge amounts of carbohydrates, sodium, and potassium chlorides, as well as small amounts of proteins and fats. An analysis of dried Irish moss showed 18.8 per cent of water; 9.4 per cent of crude protein; no fat (ether extract); 55.4 per cent of sugar, starch, etc.; 2.2 per cent of crude fiber; and 14.2 per cent of ash. A further study of digestibility of carbohydrates of the algae showed that seaweeds are not highly nutritious.

AGARDHIELLA COULTERI

Three-fourths natural size

Sarcodiotheca furcata **Red serving fork**
(Setchell and Gardner) Kylin

The red serving fork is an alga usually cast ashore on Whidbey Island and other Puget Sound areas. It is not reported from California.

The most conspicuous feature of the red serving fork is the flat, single, bright red, forked blade. A rather narrow, cylindrical stipe, arising from a small, disclike holdfast, gradually widens and flattens into the blade, which begins to fork when 1-2 inches wide. The irregular forking continues three or four times, some of the forks being much wider and longer than others. The upper ends of the wider forks are blunt, with a deep cleft in the tip of each. Along the edges of some of the forks are leaflike outgrowths. These arise in no specific order; sometimes they grow on the stipe and sometimes on the edges of the forks, while other forks have no proliferations. The over-all height of the seaweed may be 7-8 inches with a spread of 4-5 inches. The texture of the seaweed is firm and rubber-like. The fruits resulting from fertilization are scattered over the surface of the blade, appearing as prominent spherical bodies with pores at the tops. When the spherical bodies drop off, small, round holes are left in the blade.

NOTES ON SEA AND SHORE

Immense pressure and darkness are two of the governing conditions of the sea. At 200-300 feet the red rays are blotted out of the water; at 400-500 feet the greens fade out; at 1,000 feet dark blue alone remains; and below 1,500-2,000 is the blackness of the sepulcher. In some curious way the colors of marine plants and animals tend to be related to the color of the zone in which they live. Surface fish are often green or blue. At a thousand feet fish are silvery, red, or brown. At depths greater than 1,500 feet they are black, deep violet, or purple. Jellyfish, which are transparent in the upper waters, are deep brown at 1,000 feet. Plant life disappears entirely below 600 feet, and few algae live deeper than 200 feet.

SARCODIOTHECA FURCATA

Natural size

Plocamium pacificum Kylin Sea comb

The sea comb is attached to rocks partially buried in the sand in quiet waters between the mean low-tide and the -1.5-foot tide levels from British Columbia to southern California.

The dainty sea comb reaches a height of 3-8 inches, with a number of loose, narrow, subcylindrical, or compressed branches spreading nearly as wide as the height. The most noticeable features are the almost transparent, elastic texture and the 4-7 feather-like, curved branchlets. The branchlets extend from one side of the branches only, that is, from the side opposite the main axis, with the uppermost branchlet being most strongly developed. The branchlets, close set like the teeth of a comb, taper to acute points which are outwardly curved. The long, sprawling branches are arranged so irregularly on the main axis that the axis is often obscured. There is considerable variation in the density of the branching, but usually the branches near the tips are more closely placed than those nearer the base. Specimens from deep water are more loosely branched than those found higher on the beach. The holdfast is a tiny disc, or the prostrate lower branches may attach themselves to the substratum to form the holdfast. The sea comb has a rich rose hue, but sometimes the color is obscured by the sand or by being overgrown with diatoms. The sex organs of the sea comb are on separate plants. After fertilization the male cells near the tips of the branches become inflated.

NOTES ON SEA AND SHORE

Only a small per cent of the sea floor has sufficient light to support attached plants, which cannot live at depths greater than a couple of hundred feet. Even this small area is often rendered useless because of great stretches of mud, sand, or other unfavorable features for holding plants in place. For this reason the attached plants cannot support the great numbers of animals in the seas. Therefore the primary food production is largely the function of the unattached floating plants, the microscopic diatoms and dinoflagellates which occur in tremendous numbers.

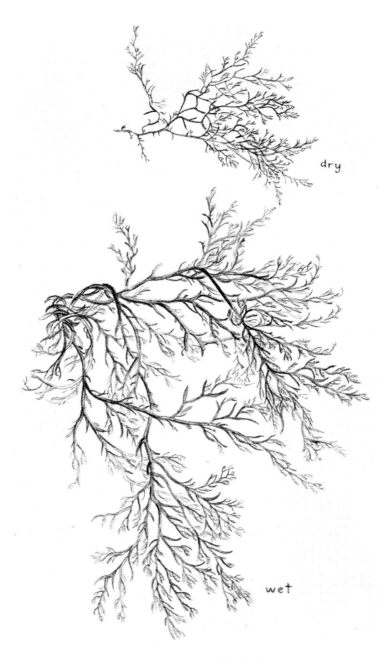

PLOCAMIUM PACIFICUM

Natural size

Gracilaria confervoides (Linnaeus) Greville   Sewing thread

The sewing thread lies half buried in sand or attaches itself to rocks at low-tide level or below. It is found in varying quantities from Vancouver Island to southern California. This is a cosmopolitan species growing throughout the world.

It is easy to identify the sewing thread by its long, naked, whiplike shoots, from 2-6 feet long and 1/32 inch wide. The shoots, which live from year to year, are cylindrical and irregularly and sparingly branched, the branching usually taking place near the base. The texture is cartilaginous, becoming horny when dry; the color varies from yellow to red-brown to purplish black; and the branches taper to sharp points at the apex.

In a study made in North Carolina, the plant was found to develop in two phases—one producing spores typical of red algae, and the other a loosely drifting plant which does not form spores. This form lies loose in shallow waters or drifts about as it grows. Reproduction is merely the accidental breaking apart of the plant. In the spore-forming plant, male and female organs are on separate thalli. After fertilization the mature fruit is visible as small nodules or bumps on the surface of the thallus. Those that develop from spores grow from disclike holdfasts composed of closely compact threads.

NOTES ON SEA AND SHORE

Gracilaria confervoides is one of the seaweeds from which agar—a solidifying agent used in foods, varnishes, sizings, etc.—is made. In North Carolina the attached stage which produces the spores does not occur in sufficient quantities to be commercially valuable, but the loose drifting phase is abundant. From August until October this accumulates in great masses along the shores where the tides are not strong enough to sweep it out to sea. Commercial collectors gather this material. After a complicated manufacturing process, blocks of frozen agar are obtained. The agar is later worked into dry, brittle sheets which are hammer milled or pulverized and made ready for the market.

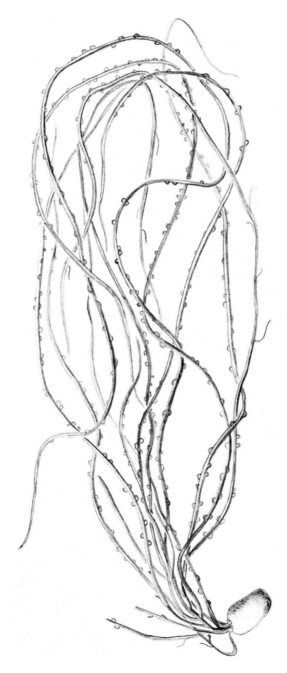

GRACILARIA CONFERVOIDES

Natural size

Ahnfeltia gigartinoides J. Agardh   Loose Ahnfelt's seaweed
Ahnfeltia plicata (Hudson) Fries   Bushy Ahnfelt's seaweed

Ahnfelt's seaweed is an abundant genus that lies half buried in the sand or grows on rocks between the 1.5- and -1.5-foot tide levels from Vancouver Island to southern California.

The outstanding feature of this genus of seaweed is the presence of naked, wiry, rigid, erect branches. In *Ahnfeltia gigartinoides* the sprawling, naked branches are 4-10 inches in height with branches 1/32-1/16 inch in diameter. The loosely arranged branches are forked as many as 10-15 times with an interval of ½-1 inch between the successive forks. The cylindrical branches are usually on one plane, and the plant is without a central axis. In *Ahnfeltia plicata* the thallus is a dense, bushy mass of more or less entangled branches 2-5 inches in height. The branches are forked 5-10 times, the terminal ones being bent like a bow. *Ahnfeltia* is a perennial plant which grows very slowly. Only one or two forks are believed to be formed each year. Occasionally outgrowths appear near the bases of the branches. The fruiting bodies form conspicuous wartlike growths nearly the thickness of the branches bearing them. Both species of *Ahnfeltia* are very deep purplish red to black. The erect thalli grow from a prostrate, branched, cylindrical rhizome.

NOTES ON SEA AND SHORE

The making of agar, the gelatinous substance obtained from certain seaweeds, is being engaged in by a number of companies in the United States. The Japanese, however, have developed the industry to the highest degree. In 1930 the Russians undertook to produce agar from *Ahnfeltia*. They learned of its possibilities from the Japanese at Sakhalin Island, north of Japan, where large quantities of *Ahnfeltia* are found. One Russian scientist, Kizevetter, claims that *Ahnfeltia* agar "equals the high grade agar made abroad and in some respects excels it." Russian scientists have published many reports on the technology, manufacturing processes, chemical nature, physical properties, and uses of Russian agar.

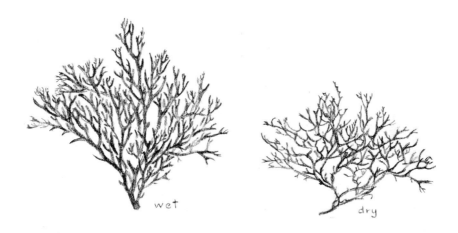

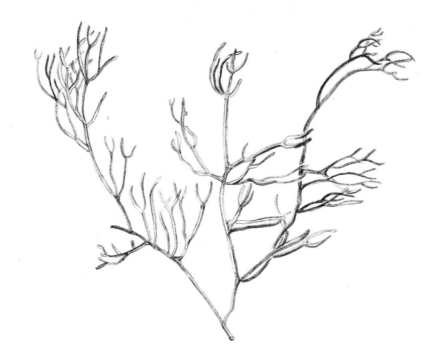

AHNFELTIA PLICATA (above)

Natural size

AHNFELTIA GIGARTINOIDES (below)

Natural size

Gigartina Stackhouse　　　　　　　　　　　　　Grapestone

Gigartina is a genus of seaweed in which the species are extremely difficult to distinguish. The species *Gigartina latissima, Gigartina unalaskensis,* and *Gigartina mamillosa* are widely distributed in the North Pacific, but no attempt will be made here to differentiate them.

Common to all species of *Gigartina* are the prominent outgrowths, resembling grape seeds, that almost cover the surface of the thallus. They are particularly dense on both sides of the upper part of the blades. Occasionally the outgrowths are so large they become bladelets. Sooner or later the thallus of all species has these bumps or outgrowths, although they may not be present during the early stages of development. The fruiting bodies are located at the base of the outgrowths. The blades of *Gigartina* are variously divided; some are entire, some forked, some leaflike, and some irregular. Likewise some species of *Gigartina* are quite large, up to 15 inches in length and 6-8 inches in width, but most species in this area are comparatively small—from 2 to 5 inches—with many divided blades which are ⅛-1 inch in width. The color of *Gigartina* is also variable; sometimes it is a clear dark red, but usually it is black with a heavy, leathery texture. Almost all species become black and shrink considerably when out of water. One or more flat blades (sometimes a number) arise from a disclike holdfast which shows zones of annual growth. Where the substratum is favorable, *Gigartina* is so abundant it forms a definite algal zone between the 2-foot and the mean low-tide levels.

NOTES ON SEA AND SHORE

The most important edible seaweed in the United States is carrageen or Irish moss. While most of the Irish moss is made from *Chondus crispus, Gigartina* has recently been used in its production. The value of seaweed colloids is due to their remarkable thickening, jelling, and stabilizing powers. Irish moss is used in making cosmetics and pharmaceuticals and as a stabilizer in prepared foods such as chocolate milk, cake icings, pie fillings, milk puddings, jellied consomme, jellied fish, and confections.

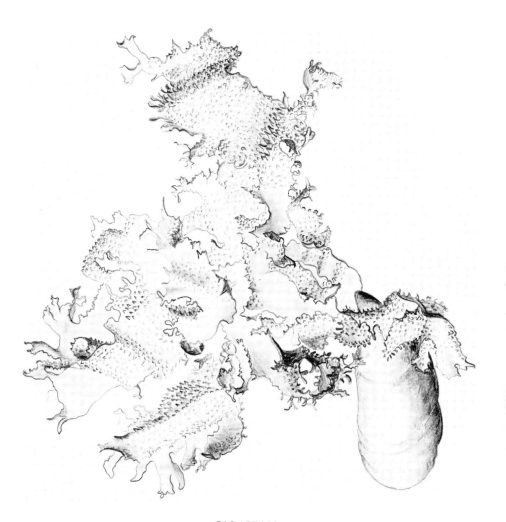

GIGARTINA

Three-fourths natural size

Gigartina exasperata Harvey and Bailey   Turkish towel

The turkish towel is a species of *Gigartina* common on Washington shores between the mean low-tide and the -1.5-foot tide levels on rocks exposed to the surf.

The turkish towel can be distinguished from other species of *Gigartina* by the single, large, stiff, irregularly shaped blade and the short, uniform outgrowths on its surface that give it the texture of a turkish towel. As in other species, these outgrowths are associated with the fruiting bodies and are particularly prominent on both the flattened faces of the upper part of the blade. On some specimens the outgrowths are so large they are like small bladelets. The alga varies widely in shape and size, but it is often 8-11 inches long or ½-3 times the breadth at the widest point. The blade arises from a short stipe around which grow several small, insignificant blades. Above the stipe the blade widens gradually for a short distance, then widens abruptly until it becomes broadly rounded at the apex. As the plant becomes old, the blade is punctured with holes and the margins are deeply cut. Also, when it is old, the blade shrivels, becoming small and thick and resembling hard, coarse rubber. This seaweed varies considerably in color depending on its age. A young specimen is yellowish red to bright red with the surface so shiny it looks as though it had been shellacked. An old specimen is dull and black. One or more blades arise from a disc-shaped holdfast.

NOTES ON SEA AND SHORE

New uses of Irish moss and its close relatives (one of which is *Gigartina*) are constantly being developed. In addition to its uses in foods it is widely used in emulsions and tablets and as a therapeutic agent for constipation, in insect sprays, water-base inks, sizing for cloth and paper, and as a clarifying agent in making beer. For centuries Irish moss has been used in preventing scurvy and goiter and valued for its antiscorbutic vitamins and large amounts of iodine, bromine, calcium, magnesium, and potassium salts.

GIGARTINA EXASPERATA

Three-fourths natural size

Iridophycus Setchell and Gardner   Iridescent seaweed

A number of species of *Iridophycus* are found midway up the beach attached to rocks exposed to heavy surf. The species are difficult to distinguish.

The most striking feature of the iridescent seaweed, as its name implies, is its iridescence. When in the water *Iridophycus* is brilliantly colored. As the light falls from different directions, it exhibits a succession of bright metallic hues of blue, green, and purple. It is also characterized by its single large, flat blade, its rather thin, rubber-like texture, and its bright red to dark purple color. Sizes and shapes for this seaweed vary from specimens whose thallus is 3-4 inches in length to those with a length of 6 feet, depending on the species. Usually the blade is simple and smooth with an acute apex, a rounded base, and a short, cylindrical stipe. There is usually but one blade, but in some species several small, undeveloped blades arise from the holdfast. The edges of the blade are frequently cut and ruffled. The male and female organs are on separate thalli, with the fruiting bodies appearing as clusters of very small dots scattered over the surface of the blade. A disc-shaped holdfast anchors the plant to the substratum.

NOTES ON SEA AND SHORE

The Hawaiians, like the Japanese, enjoy eating seaweeds. Until very recently they gathered and ate tons of seaweed daily and also imported large quantities from the Orient. Even today seaweeds constitute an important part of the vegetable diet of the poorer Hawaiians. They use approximately 70 species of seaweed, 40 species regularly and 30 more occasionally. The seaweed is preserved by salting and by placing it between layers of ti leaves (any of several species of Asiatic trees or shrubs), which prevent the seaweed from drying out. It may also be preserved in weak brine. However, the greater part is washed and eaten raw as a relish with poi, meat, or fish.

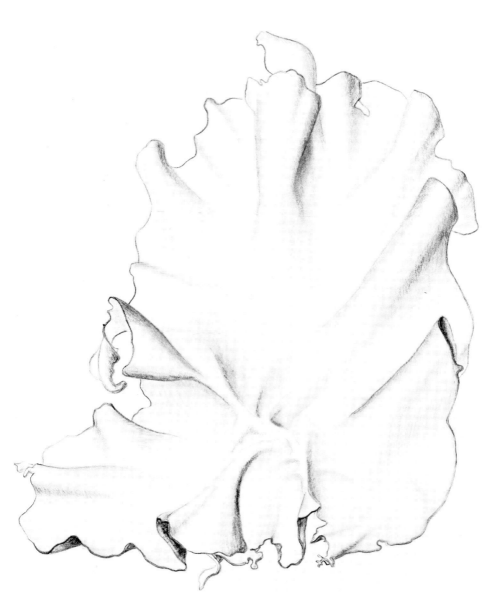

IRIDOPHYCUS

One-half natural size

Halosaccion glandiforme (Gmelin) Ruprecht   Sea sac

A seaweed which ranges from western Alaska (the Aleutian Islands) to Mexico is the sea sac. It grows abundantly in the mid-tide range of 2.5- to 1-foot tide levels in rather restricted areas. In areas where sea sacs are present there may be so many specimens that they form a conspicuous belt about 1-1½ feet wide. Above or below this zone there are few sea sacs.

One of the easiest seaweeds to identify is the sea sac, an erect, hollow alga standing upright on a short, cylindrical stipe. It is approximately the length and circumference of a human finger, about 4 inches long, ½-¾ inch in diameter, and rounded at the top. The walls of the sac are thin and smooth with the texture of rubber gloves. When the plant is young, the sac is filled with water and has microscopic openings at the tip so that when the sac is compressed water spurts out in fine jets. In older specimens the sac is somewhat compressed, but it is still filled with water. Eventually the walls thicken, the tip becomes eroded, and the sac is filled with sand. Frequently several sacs are borne on a single holdfast, all but one sac being small and undeveloped. Sea sacs often grow very close together, as a protection against injury by wave action. The sea sac varies from yellowish brown to olive-green to reddish purple, the color being determined by the age of the plant.

NOTES ON SEA AND SHORE

Certain seaweeds thrive in definite zones on the beach. Apparently the conditions at various levels favor one type of seaweed above all others. The plants in the highest zone live above high-tide mark and are only splashed with water, while others live in such deep water that they are seen only when the tide is unusually low or when they are torn from their holdfasts and swept ashore by storms. Of course the greater number, such as the *Halosaccion*, are exposed to periods of submersion and exposure each day. The marked zonation of seaweeds in the littoral region is no doubt a result of progressive exposure factors.

HALOSACCION GLANDIFORME

Two-thirds natural size

Rhodymenia palmata forma mollis Setchell and Gardner

Dulse or red kale

Dulse or red kale is a widely distributed seaweed found in tide pools or on rocks at mean low-tide mark, where it often forms a distinct zone, from Alaska to central California. It also attaches itself to blades of *Nereocystis* (ribbon kelp).

Dulse may be recognized by the dull red, forked, irregularly divided, rosette-like blades which have the texture of thin rubber. The numerous blades, 5-15 inches tall and 1-3 inches wide, arise from a short, inconspicuous stipe. The blades are flat, membraneous, and variable in shape, sometimes oval or wedge-shaped, sometimes cut into a few segments or into narrow ribbons. The margins of the segments, especially those near the base, have numerous outgrowths, but these usually remain small. The blades are on one plane and are without midrib or markings of any kind. Gall-like growths or irregular swellings sometimes found on dulse are due to the presence of copepods, tiny animals which attach themselves to the alga. An abnormality of the life cycle of *Rhodymenia palmata* is indicated by the apparent nonexistence of female plants. The male organs are located in small, irregularly shaped clusters on the flattened faces of the blades. The scientific name *(Rhodymenia palmata)* indicates that the seaweed is shaped like the palm of a hand. The attachment to the substratum may be either by a small disc or by a number of rootlike hairs or rhizomes.

NOTES ON SEA AND SHORE

Dulse is a widely used seaweed. In many parts of the world it is used as food, as a relish, or as a medicine. It is eaten raw, chewed like gum, eaten with fish and butter, or boiled with milk to which rye flour has been added. During the centuries when famine was rife in Ireland, dulse and potatoes formed the staple foods for the people along the coasts. In Mediterranean countries today the people use dulse in ragouts and other prepared dishes. It gives the dish a pleasing red color, and the gelatinous substances in the tissues thicken the food. Dulse is also used in many parts of the United States. Some is gathered on New England shores, but it is usually brought from the Canadian provinces for sale in the markets. Dulse is still gathered and used as a relish by the Indians (and seaweed enthusiasts) in the Puget Sound region.

RHODYMENIA PALMATA

Two thirds natural size

Rhodymenia pertusa   **Red eyelet silk**
(Postels and Ruprecht) J. Agardh

Red eyelet silk is usually found near or below low-tide mark, but it may also be dredged from depths of 30-50 feet. It is a characteristic subarctic, sublittoral species, not common on Pacific Coast beaches.

It is a striking seaweed with a large, single, flat, bright red thallus dotted with holes of more or less uniform size which suggests the common name, red eyelet silk. A short stipe arises from a minute, disclike holdfast. From this the blade—which when fully grown may be 12-20 inches long and 5-10 inches wide—begins to broaden almost at once. It soon reaches its maximum width, which it maintains, the apex being wide and rounded. Usually the only markings on the blade are the perforations. The perforations appear on any part of the blade, new ones forming among the older ones. Their formation seems to be accompanied by the destruction of the tissue. Occasionally the blade has leafy shoots extending from the margin. When young these are rounded bumps, but later they become miniature blades. The blade is as thin as tissue paper.

NOTES ON SEA AND SHORE

The brilliant and changing color of the ocean provides much of its beauty and fascination. As we know, the color of the sea varies from deep blue in the open ocean to an intense green in coastal waters, while inshore waters may be brown or red. Several theories as to the causes of the color of the ocean have been suggested, but most researchers now agree that the blue color is a result of the amount of radiation scattered against the water molecules themselves or against minute suspended particles smaller than the shortest visible wave lengths. The transition from blue to green is probably the result of water-soluble yellow pigments in sea water. The combination of the yellow pigment and the blue of the water leads to a scale of green colors. In coastal areas the water appears brown when large quantities of suspended organic or inorganic matter are present or when mineral particles are carried into the sea after heavy rainfall. The red color is due to a concentration of millions of one-celled algae or dinoflagellates.

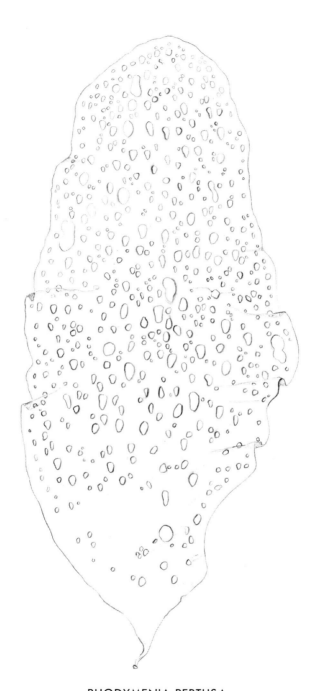

RHODYMENIA PERTUSA

One-third natural size

Gastroclonium coulteri (Harvey) Kylin   Sea belly

The sea belly is a neat-looking seaweed adhering to rocks lying between the 2- and the -1.5-foot tide levels from Puget Sound to southern California.

The sea belly has erect, cylindrical, sparsely arranged, forked branches, 5-8 inches high and 1/32 inch wide, which arise from a broad, heavy, disc-shaped holdfast. Some of the major branches fork once or twice. The lower part of the branch is usually naked, while the upper part has a number of irregularly arranged, short, naked, lateral branchlets. The lateral branchlets are ¼-½ inch long, club-shaped, and broadly rounded at the tips. Transverse partitions constrict the hollow branchlets at regular intervals. However, the lower part of the major branches is solid and is not constricted. Usually the sea belly is a dark olive-brown, sometimes a dark green. The male and female organs are on separate thalli, with the male organs situated in irregularly shaped patches on the bladder-like terminal branches. After fertilization the fruit stands above the thallus and is surrounded by an urn-shaped envelope with a wide mouth. Several species of *Gastroclonium* found in warm seas are beautifully iridescent.

NOTES ON SEA AND SHORE

The sea contains many chemical elements, but none has stirred man's dreams more than gold. Gold is in all the waters of the world, enough of it to make every man, woman, and child in the world a millionaire. But as yet no inexpensive, practical method of extracting it has been discovered. After the First World War, a German scientist dreamed of paying off the German war debt by robbing the sea of its gold. He equipped a ship with a laboratory and an infiltration plant, but the quantity of gold secured was disappointing and the cost exorbitant. It is estimated that in a cubic mile of sea water there is $93,000,000 in gold and $8,500,000 in silver. At present only the corals, sponges, and clams have a feasible method of extracting these valuable materials.

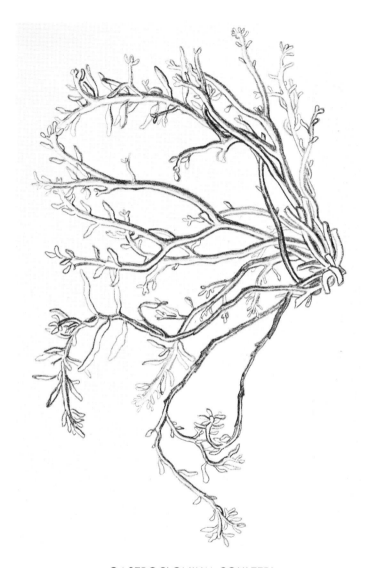

GASTROCLONIUM COULTERI

Natural size

Antithamnion pacificum (Harvey) Kylin   **Hooked skein**
Antithamnion uncinatum Gardner   **Hooked rope**

The hooked skein and the hooked rope are attached in large masses to the stipes and air bladders of *Nereocystis* (ribbon kelp) and *Macrocystis* (bladder kelp). They are found wherever these kelps are washed ashore from Alaska to southern California.

The distinguishing features of these seaweeds are their attachment in conspicuous masses to the big kelps, their bright colors, and their delicacy. In both species the individual branches and branchlets are as fine as the most delicate thread, but when these are massed together the hooked rope gives the impression of a frayed rope, while the hooked skein looks like a skein of thin silk floating on the water. In the hooked rope the individual thalli are 5-7 inches long; in the hooked skein they are 3-4 inches long. Both species are a lovely, uniform, dark red. When looked at through a microscope, the hooked skein is seen to have an obscure axis bearing a number of major branches, with a branchlet opposite each major branch. The branchlets are simple and curve upward toward the top of the branch bearing them. In the hooked rope there is a prominent axis, but this is often obscured by many long branches. Each major branch has a branchlet opposite the long branch. Some of the branches in the hooked rope elongate, and the tips turn downward and hook around other branches, giving the rope appearance. This species is somewhat firmer than the hooked skein. In both species the male and female organs are on separate thalli. The holdfasts of these seaweeds may be discs, or the cells may join together end to end, forming creeping threads which penetrate the host.

NOTES ON SEA AND SHORE

Perhaps no other group of plants exhibits so many different colors as does that of the algae. The common names—brown, green, red, blue-green—give no clue to the possible color variations within the groups, for they exhibit every hue within the color range. Not much is known about the chemistry of color in seaweeds.

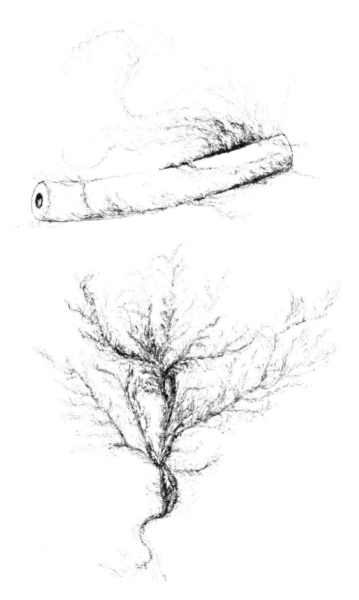

ANTITHAMNION PACIFICUM on NEREOCYSTIS (above)

Natural size

ANTITHAMNION UNCINATUM (below)

Natural size

Callithamnion pikeanum Harvey   Beauty bush

From southern Alaska to central California, the beauty bush grows on the faces of rocks and on other seaweed between the 3.5- and 1.5-foot tide levels. It is abundant everywhere.

The beauty bush has many erect, alternate fronds borne on strong, cylindrical axes. The branches are arranged in densely crowded spirals on all sides of the axis. In fact, they are so crowded that it is difficult to distinguish the individual thalli. Extending from the main branches are very short (1-2 mm.) recurved filaments. The main branches, as well as the branches of the second and third orders, are covered from base to apex with a thin, barklike covering. A microscope is needed to see this covering. All the branches are curved inward, with the ultimate branches sharply pointed. As a general rule the beauty bush is 3-7 inches tall, occasionally 10-12 inches, with the primary branches 1-3 inches long; these in turn bear extremely short second- or third-order branches. The spores are situated on the adaxial side of the ultimate branches. The entire seaweed is a uniform dark purplish brown. The holdfast is a rhizoid, formed by filaments (rows of cells end to end) which emerge from the basal cells. These sometimes unite to form a dense mass. The holdfast may either attach itself superficially to the substratum or penetrate it. The beauty bush lives for several years.

NOTES ON SEA AND SHORE

Most red algae are characterized by comparatively elaborate thalli, usually red because of the presence in the cells of phycoerythrin and phycocyanin. The red algae are distributed in all seas, but they are most abundant in Australasia and the warmer regions, where many highly specialized forms occur. They are partial to the lower sublittoral or intertidal zones, although a few can endure considerable exposure above high-tide mark. The red algae include two subclasses—the Bangioideae, a very small group of simple types, and the Florideae, to which belong most of the red algae.

CALLITHAMNION PIKEANUM

Natural size

Griffithsia pacifica Kylin　　　　　　　　Griffith seaweed

The Griffith seaweed grows on rocks between the 0.5- and the 1.5-foot tide levels, although it is sometimes dredged from depths of 30-40 feet. It is not a common seaweed, but it is found occasionally on the Pacific Coast from Puget Sound to southern California.

The Griffith seaweed is a small, clear, reddish-pink, threadlike species with branches which spread out into a fanlike pattern. The branches, 2 inches long and 0.5-1 mm. wide, remain distinct from one another, never matting together. The branches are naked, without branchlets or outgrowths of any kind. Usually the branches are regularly forked to the third or fourth degree and are somewhat broader at the top than at the base. The cells are so large that they can be seen with the naked eye. A gelatinous envelope encases the entire thallus. Male and female organs are on different thalli. However, no matter at what season of the year collections are made, fruiting specimens are extremely difficult to find. The texture of the Griffith seaweed is rather harsh. Strong rhizoids which develop from the basal threads fasten this seaweed to the substratum. The rhizoids curve around the lower branches and thus give rise to new shoots.

NOTES ON SEA AND SHORE

Red seaweeds heal their wounds and regenerate themselves by several different methods. A red alga can regenerate from a small fragment of the thallus or, as in Griffith seaweed, from an isolated cell. Likewise, if fragments of the thallus carry active reproductive materials, new thalli may be produced from them. In an alga which has a compact thallus, wounds are healed by active division of uninjured cells. Damage to the apex of an alga results in an outgrowth of the central cell into a forked apex. In plants whose tops have been injured, branches arise from other than the usual places. In some genera, if the axis is broken or damaged near the base, one of the basal shoots takes its place, whereas, if the injury takes place in the upper part of an older plant, laterals grow into long shoots which give the plant a bushy appearance. In *Antithamnion*, damaged cell walls increase as much as ten times in thickness.

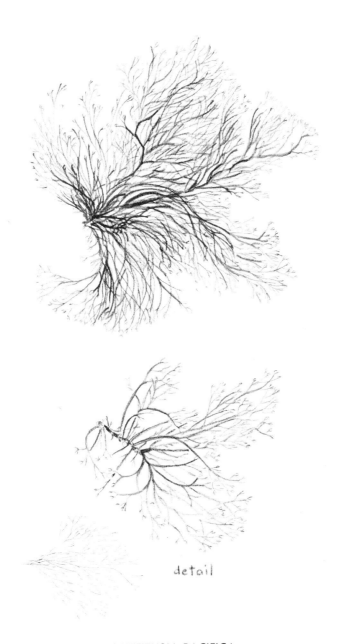

GRIFFITHSIA PACIFICA

Natural size

Ceramium pacificum (Collins) Kylin   Pottery seaweed

The pottery seaweed is a widely distributed alga attached to shells or rocks or epiphytic on other seaweed either in tide pools or in deeper water. It is found in almost every part of the world.

There are a number of species of *Ceramium* on the West Coast of the United States, but the differences are so small that it is difficult to distinguish them. Even the color varies from a clear dark red to brown or purple, depending upon the habitat. Probably the most noticeable features of this genus are the transverse bandings on the younger parts of the plant and the widely diverging hairlike branches, which are forked 5-6 times. The irregular forkings are closer together near the tip of the frond than near the base. The plant grows in sprawling, loosely arranged tufts 4-6 inches high, with the branches as fine as hairs. The sprawling branches, diverging from 30-60 degrees, are all about the same length, making the plant level-topped. On the branches are numerous short, threadlike branchlets ending in hooked spines. Short, stiff, hairlike growths extend along most of the branches. An unusual feature of the genus is the tendency of the tips to bend toward each other, the uppermost incurved like a pair of pincers. The pottery seaweed is fragile and dainty, and the point of attachment is almost invisible. The seaweed often floats in upon the waves, detached from its holdfast.

NOTES ON SEA AND SHORE

Ceramium, like a number of other seaweeds, possesses sickle-shaped hooks which, upon coming in contact with another branch of the same or a closely related alga, coil around it like tendrils. In the species which do this, the curved tips are thicker than the lower part of the branch. In certain red algae the inrolling of the tendrils and the production of rhizoids are the result of definite contact stimuli. If an alga of another species is encircled, there is no response, but when a tendril grasps a branch of the same alga, both structures respond and form rhizoids.

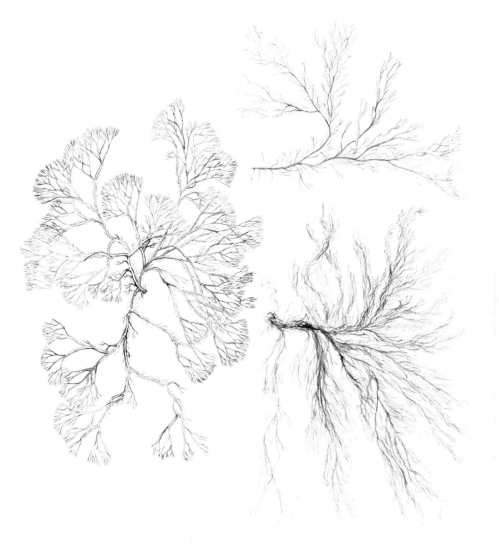

CERAMIUM

Three-fourths natural size

Microcladia coulteri Harvey Delicate sea lace

The delicate sea lace is found from Alaska to southern California. It lives on other algae which inhabit the lower reaches of the tide and extend a short distance beyond the low-tide mark. It is quite abundant and easily recognized, for it contrasts sharply with the host plant.

The sea lace is a deep rose color. It is epiphytic especially on *Gigartina*, *Prionitis*, and *Grateloupia*, to which it attaches itself so firmly that it is difficult to remove without tearing both host and epiphyte. As the common name suggests, the alga is delicate and lacy. An average-sized specimen measures 4-10 inches in height, with several orders of rather firm, narrow branches extending from a central cylindrical or slightly compressed axis. The lower third of the plant has primary branches up to an inch long; those of the higher orders become progressively shorter until those of the final orders are only ⅛ inch long. This gives the plant a pyramidal shape. The regular alternate branches arise from both sides of the axis, with all the branches lying on one plane. The branches of the first, second, and third orders are usually straight; those above are curved and forked like a pair of pincers. The fruiting bodies are densely crowded on the final three orders of branches.

NOTES ON SEA AND SHORE

The many epiphytic red seaweeds are for the most part attached only superficially to their hosts. A certain number, however, show a more intimate relationship by penetrating so deeply into the host that they cause some destruction to the surrounding cells. There is little evidence of real parasitism in red algae, and the nourishment they take from the host is small, if any. It is noteworthy that, in the more definite examples of parasitism among red algae, host and parasite usually belong to the same order of plants, often to the same family. This specialization is striking in view of the dense communities of seaweed from which a choice of hosts is available.

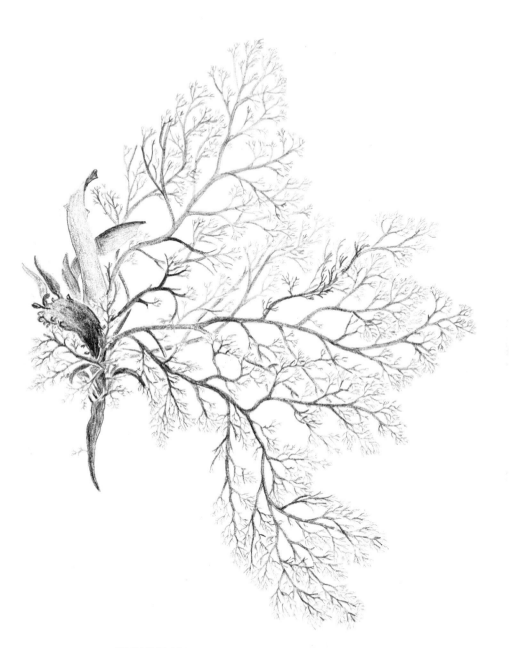

MICROCLADIA COULTERI on PRIONITIS LYALLII

Natural size

Microcladia borealis Ruprecht   Coarse sea lace

The coarse sea lace grows on the face of a rock more or less exposed to the surf between the 2.5- and the 0.5-foot tide levels. It is abundant on most beaches from Alaska to central California. This species of *Microcladia* differs from *Microcladia coulteri* by being attached to rocks rather than to other seaweeds. The coarse sea lace is a graceful red alga with a number of short, stiff, cylindrical to slightly compressed branches arising on one plane only. The erect branches, each of which is divided into 5-6 orders of smaller branches, are generally bare for the lower ⅓ of their length. In each of the orders the branches are arranged alternately and are on one side of the branch only, like the teeth of a comb. The branches of the first and second orders are bent like bows; the branches of the higher orders are forked like a pair of pincers. There is much variation in the size of the coarse sea lace, but those found on the Pacific Coast average from 4 to 6 inches in height and are almost hairlike in width. Usually this species is a deep red, but occasionally it is a dark olive-gray. It is attached to the substratum by a prostrate, rhizome-like system of branches. The male and female bodies are on separate thalli. The spore cases, which divide to form four spores, are densely crowded on somewhat inflated branch tips.

NOTES ON SEA AND SHORE

The distribution of seaweeds is brought about by many factors. Ocean currents have a powerful influence upon their distribution by extending the range of the spores of many seaweeds. The currents also convey the plants themselves to new localities along the shore by crossing channels, gulfs, or even seas. Certain marine algae have been introduced into new geographical areas on ship bottoms. Storms carry masses of seaweed hundreds of miles from where they were torn loose from their holdfasts. Many of the seaweeds thus transported bear live reproductive materials which later mature and form new communities. Birds also play a part in the distribution of seaweeds by carrying spores in their feathers or in their digestive tracts.

MICROCLADIA BOREALIS

Natural size

Ptilota filicina (Farlow) J. Agardh Red wing

The red wing is found from Alaska to central California between the mean low and the -1.5-foot tide levels. Often it is epiphytic on other algae, especially on coralline algae. In turn it sometimes has *Microcladia coulteri* (another epiphytic alga) growing on it.

Probably the strong central axis, 4-10 inches long with closely set alternate branches which become 2-3 inches long, is the most conspicuous feature of this alga. Its branches are compressed, harsh, brittle, and covered with a cortex. Frequently several erect shoots grow from a common disc-shaped base. The short stipe develops into a thick, coarse axis which extends the entire length of the plant with branches set close together along both sides of it. Branching is opposite, but the branch on one side is much longer than the one on the opposite side, the short one sometimes being merely a stump and difficult to see. The long and short branch of each successive pair alternate with the corresponding branches in the pair below. There are usually two orders of branching. From the edges of the branches, leaflets 1/16 inch long arise. These also have a long and short pair alternating with a short and long pair. The leaflets are sickle-shaped with both edges toothed like a saw, the teeth turning toward the apex. This seaweed is a deep, bright red.

NOTES ON SEA AND SHORE

Many species of seaweed cannot live above or below definite tidal limits. One of these limits is determined by the light penetration. Measurements of light penetration show that 25 per cent of the light is cut off at the surface and another 20 per cent at 3 feet below the surface. The light is also screened according to quality or wave length. An alga growing at a depth of 24 feet receives 10 per cent of violet-blue rays and only 1 per cent of the red rays of light. There is also a gradation in the period of exposure to light during the low tide and in the depth of submergence during high tide. In the sublittoral region it is both quality and quantity of light which determine the zonation of the seaweed population.

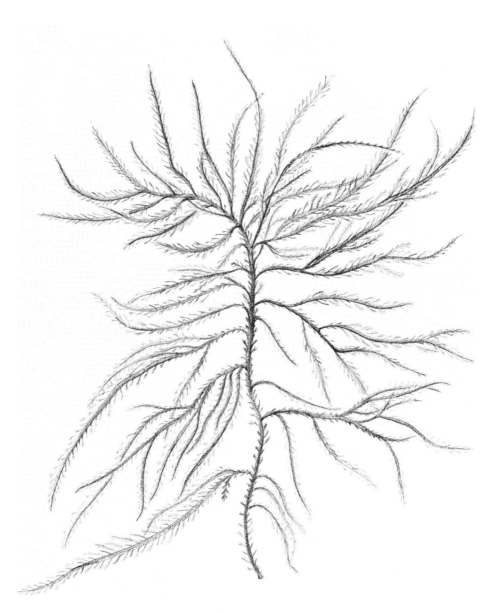

PTILOTA FILICINA

Three-fourths natural size

Delesseria decipiens J. Agardh   Baron Delessert

The Baron Delessert grows on rocks between the 0.5- and -1.5-foot tide levels from Alaska to central California. This widely distributed seaweed is partial to deep-shaded pools, shallow water, and warm waters.

The Baron Delessert seaweed is easily recognized by its delicate, graceful undulations as it sways with the movement of the water; its prominent axis and midrib; its unusual length; its bright red or purple color; and its narrow, one-celled ruffle, extending the length of the cartilaginous axis. The entire plant may be 10-30 inches in length, possibly longer. Arising from the axis are a number of alternate branches 4-5 inches long, each with a midrib similar to the main axis. These branch again into three or four orders, the midrib being prominent in all orders. Along the edges of all the branches are linear leaflets ¼-½ inch long. The one-celled ruffle, about ½ inch in width, is not only attached along the main axis but present along the length of the midrib and axis of all orders of branching. There are minute, parallel veins on the lateral ruffle and leaflets. Only the midrib and veins have more than a one-celled structure. Often the lower branches are eroded and worn away so that the axis resembles a stipe. This seaweed is attached to the substratum by a minute disc. The male sex organs are in patches irregularly scattered over the ultimate branchlets.

NOTES ON SEA AND SHORE

Most of the species of *Delesseria* are perennials living below low-water mark in the warm seas of the Southern Hemisphere. The majority of the red algae of the temperate latitudes are found either in deep water or in rock pools, for they are unable to endure exposure to the sun or to be out of water even for short periods. Several investigators have found that the red algae of the littoral zones do not show as great photosynthetic efficiency in green light as do the forms from deeper water. The deep-water forms apparently have a higher capacity for using available light.

DELESSERIA DECIPIENS

Two-thirds natural size

Erythroglossum intermedium (J. Agardh) Kylin

Dainty leaf

The dainty leaf is a small, red alga that grows on the vertical face of rocks between the mean low-tide and the -1.5-foot tide levels on Vancouver Island and the Puget Sound region; two other species, *E. Californicum* and *E. divaricatum*, are found on the Monterey Peninsula and at Santa Barbara.

The dainty leaf is not more than 2 inches tall, with several linear branches, 1/16 to ¼ inch wide, arising from the base. Some plants are wholly erect while others have a prostrate base with some of the branches becoming erect. The flat blades are irregularly branched once; the new blades grow at irregular intervals from the margins of the old ones. The edges of the blades are smooth. In this species the midrib is indistinct, as is the one-celled (monostromatic) portion lateral to the midrib. The margin of the monostromatic portion is also smooth. The branches are considerably wider at the top than at the stipelike base. At the apex of the blade is a conspicuous cell which initiates growth. The mature fertilized cells are scattered over the outer edges of the blade. These cells are surrounded by envelopes with openings at the tops. The dainty leaf is a bright rose-red color. In many ways it resembles the smaller species of *Delesseria*, for they belong to the same family.

NOTES ON SEA AND SHORE

The rapidly increasing population of the world may lead to a shortage of food for human consumption, especially of fats, oils, and proteins. Organizations like the Carnegie Institution of Washington, Department of Plant Biology, are already making efforts to solve this problem by investigating the possibility of using microscopic algae as a source of food. The work there has centered largely around the idea that the algae, like other plants containing chlorophyll, are able to convert inorganic compounds into organic matter by means of light energy through photosynthesis. Experiments have shown that growth of the microscopic alga, *Chlorella,* a fresh-water or soil alga, can be controlled to produce cells of either a high-fat or a high-protein content. These show astonishing possibilities for human use.

ERYTHROGLOSSUM INTERMEDIUM

Twice natural size

Polyneura latissima (Harvey) Kylin   Crisscross network

The crisscross network is found between the mean low-tide and the -1.5-foot tide levels on the vertical faces of rocks sheltered from the full force of the surf from Vancouver Island to Lower California. It is also dredged from depths of 30-35 feet.

Probably the most conspicuous feature of the crisscross network is the extremely thin blade on one plane which is covered, except on the margins, with a network of veins or nerves. The alga is usually 3-5 inches tall, but it may reach a height of 10 inches and is half as wide as tall. The one or more blades, borne on inch-long stipes, are wedge-shaped or oval and are undivided when young but become deeply cut or frayed when older. Younger portions of the blade are made up of only one cell layer, while older portions have more than one layer. The cross connections or veins are visible in the older, lower part of the blade but inconspicuous in the younger parts. Occasionally the blades have small outgrowths extending from the margins. A lovely deep red or rose color makes this tissue-paper-thin seaweed very attractive. In fact, its texture is so thin and fragile that it crumbles easily. The seaweed secures itself to a rock either by a disc-shaped holdfast or by flattened, irregularly divided, ribbon-like branches. The male sex cells appear as tiny dots scattered over both surfaces of the blades, particularly in the areas between the veins.

NOTES ON SEA AND SHORE

Although abundant in all latitudes, the greatest number of red algae are met in the warmer seas. Yet, strange to say, they reach their most luxurious growth in the polar seas. Near the poles red algae may attain a length of 5 or 6 feet, but in temperate and tropical regions most of them are only a few inches in length. Many of the delicate species are among the most beautiful objects in the sea. Certainly from the standpoint of color they are beautiful and varied, sometimes being red, sometimes purple, violet, or even brown or green. The deeper-growing species are the more purely red. Many of them are iridescent.

POLYNEURA LATISSIMA

Natural size

Hymenena flabelligera (J. Agardh) Kylin   Veined fan

The veined fan is exposed on the vertical face of rocks between the 1- and the -1-foot tide levels where the surf beats hard from Puget Sound to central California.

This seaweed may be confused with others whose exterior appearance is similar, notably *Callophyllis edentata*, yet their development and structure are quite different. In the veined fan some of the thalli are prostrate, and some are erect, arising from a broad disc. The erect branches are on one plane, fan-shaped, divided into long segments with rounded tips and smooth margins. The thallus is 5-10 inches high with inch-wide segments which take on the general shape of a fan. The blade is tissue-paper thin, and the color ranges from deep salmon pink to dark red, verging on purple. At the lower end the segments have distinct midribs. These soon disappear, and the upper part of the segments is covered with a rather complex system of tiny, fan-shaped veins. The margins of the blade and the midrib are made up of more than one cell layer, while the center part has but one cell layer. The male and female organs appear on separate thalli. Spores are in narrow patches running lengthwise on the upper part of the blade segments. These, too, are arranged in a fanlike pattern.

NOTES ON SEA AND SHORE

We all know that water is essential to the maintenance of life, for it constitutes 80 per cent or more of all active protoplasm. It is the most efficient of all solvents and carries in solution the natural gases, oxygen and carbon dioxide, as well as the mineral substances necessary to the growth of plants and animals, and it is an essential raw material in the manufacture of plant foods. The water also offers the most intimate relationship between itself and the animals and plants which live in it. Because of the stability of the physical conditions of the water and its composition and concentration of dissolved salts, the organisms in the sea have not been forced to develop highly specialized systems to protect themselves against sudden and intense environmental changes.

HYMENENA FLABELLIGERA

Natural size

Cryptopleura ruprechtiana (J. Agardh) Kylin

Ruche or hidden rib

The ruche or hidden rib is a seaweed attached to rocks between the 0.5- and the -1.5-foot tide levels from British Columbia to southern California.

The most conspicuous features of this seaweed are the flat thallus divided like an open fan, the network of veins and ribs that cover the surface of the blade, and the presence of densely crowded outgrowths on the edges of the blade. The outgrowths look like a tiny ruching or ruffle, 1/16 inch wide. Arising almost directly from the holdfast, the thallus, with an inconspicuous midrib broadens rapidly and begins almost at once to fork into a number of irregular, ribbon-like, widely diverging segments. A typical specimen may be 6-8 inches wide and 7-9 inches high. Primary segments are cut nearly to the base, the secondary ones less than ⅓ the distance to the base. The lower segments may be a couple of inches wide, the upper ones less than half an inch. The tips of the segments are slightly lobed and are without the ruffled outgrowths. This seaweed is a beautiful, bright, purplish red with a tissue-paper-thin texture. It has a narrow, branched, ribbon-like holdfast formed by the ends of certain branches which bend upward and grow erect. The male and female sex organs are on separate thalli, the male cells lying in patches along the margins of the blade segments or in the ruching.

NOTES ON SEA AND SHORE

In laboratories all over the world scientists are experimenting with phytoplankton (microscopic floating plants) as a possible source of food for man. This amazing material contains protein, fat, starch, vitamins—in fact, every component needed to sustain life. Many overpopulated countries of the world, notably Thailand, Japan, and Israel, are looking toward plankton to supply the protein necessary for human needs. Thailand is harvesting 5,000 tons of plankton a year from nearby seas. The processed plankton made in Thailand looks and tastes much like anchovy paste. The question of taste, however, is not an important problem because artificial flavorings can be added to make the product appetizing.

CRYPTOPLEURA RUPRECHTIANA

Three-fourths natural size

Dasyopsis plumosa
(Harvey and Bailey) Schmitz

Plumed chenille

The beautiful plumed chenille is attached to wharf pilings, washed ashore, or may be dredged from depths of 30-35 feet. Although it is present from Friday Harbor, Washington, to central California, it is more often found in deep water in tropical seas.

The most striking characteristics of the plumed chenille are the soft, chenille-like texture and the clear, bright, red color. When dredged and unbroken, this alga may be 2 feet long with branches 1 foot wide at the lower end. Toward the top the branches gradually become shorter and shorter, so that the whole is triangular. Branches are on one plane, and the alga has a strong central axis 2.5 mm. in width. The branches are alternate and irregularly spced, with short, undeveloped branches arising between the longer ones. Similarly, secondary branches appear at irregular intervals on the primary branches, with the short undeveloped branches between these. The branches are slightly flattened and densely fringed with extremely short filaments (chenille). Characteristic of *Dasyopsis* is sympodial growth, that is, a branch continues growth in the direction of the axis, and the axis continues growth as a lateral branch. Since the specimens on the beach are usually broken off, the holdfast is not visible, but no doubt it is a small disc.

NOTES ON SEA AND SHORE

Throughout the world there are perhaps 70 species of *Dasyopsis*, most of them found in tropical waters. All species are delicate and lacy. The few species found in the north Pacific are present throughout the year. Because *Dasyopsis* is handsome and delicate, it is frequently used for decorating place cards or correspondence cards or is placed under glass on serving trays. It retains its bright red color when dry, adheres closely to mounting paper, and is so delicate and lacy that it looks like an etching. In writings of a century ago, *Dasyopsis* is said to have been "known to lady collectors by the name of chenille."

DASYOPSIS PLUMOSA

Natural size

Polysiphonia pacifica Hollenberg Polly Pacific

Polly Pacific and closely related species of seaweed grow on rocks between the 2- and the -1.5-foot tide levels from Alaska to central California.

The distinguishing feature of Polly Pacific is the presence of dense clusters of delicate plants which grow from a wide, rhizoidal base. In its central range this alga may be 5-6 inches tall, with branches so narrow and matted that the individual branches are scarcely discernible. The texture is soft and lax, and the color reddish brown to black. The matted clusters, still attached to the holdfast, often break off from the substratum and float near the surface of the water. As the scientific name —Polysiphonia—indicates, Polly Pacific has many microscopic tubes or siphons extending through the length of the plant. Through a microscope, the thallus can be seen to consist of a main siphon with a band of 4-20 siphons surrounding it in a tierlike structure. The surrounding cells (pericentral) are usually parallel to the main axis. With the unaided eye one can see that the erect shoots have either obscure or distinct axes, but they soon become alternately branched, the branches giving off a great many tiny, alternate, secondary branches. The secondary branches are shed in the fall, and the plant is reclothed in the spring.

NOTES ON SEA AND SHORE

Today, in his search for wealth, man makes use of the plant and animal materials which were in the sea millions of years ago—oil pressed from the bodies of seaweeds, fishes, and other forms of life stored away in the rocks or in the layers of mineral substances. This is particularly true of man's search for petroleum, for of all legacies of the ancient seas petroleum is the most valuable. No one knows exactly how these precious pools of liquid deep in the earth were produced. But we do know that the origin of petroleum is most likely to be found in the bodies of plants and animals buried under fine-grained sediments of former seas and there subjected to slow decomposition.

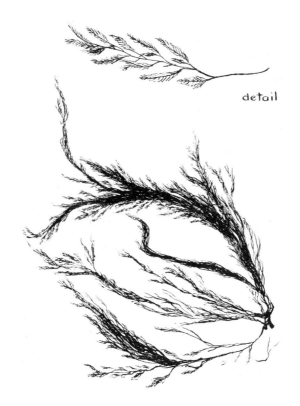

POLYSIPHONIA PACIFICA

Natural size

Polysiphonia collinsii **Hollenberg** **Polly Collins**
Polysiphonia hendryi **Gardner** **Polly Hendry**

Two polysiphonias (many-tubed seaweed), *P. collinsii,* which grows on rocks, and *P. hendryi,* which grows on other seaweed, range from British Columbia to Lower California in the midlittoral zone, roughly from the 3.5- to the -1.5-foot tide levels.

Like all species of *Polysiphonia,* these two are characterized by many siphons, by cells arranged in tiers with 4-24 cells surrounding a central layer, and by the unlimited growth of the erect branches. Polly Hendry has an extensive system of prostrate branches. The erect branches are radially branched and cylindrical, and the branch tips are turned outward. Polly Collins also has prostrate branches which become part of the rhizoids, while the erect shoots have a definite but extremely narrow main axis. The erect branches, an inch long, hairlike in width, and on one side of the axis only, are also radially branched and cylindrical. The plants of both species usually grow in reddish-brown matted tufts so heavily encrusted with diatoms that one can scarcely see the plant itself. These seaweeds either are attached by one-celled rhizoids extending from the base of the erect branches or arise from prostrate basal filaments. Great numbers of thalli grow from the rhizoids.

NOTES ON SEA AND SHORE

Although seaweeds are less diverse than land plants, there are almost endless variations in pattern, size, and complexity of cell arrangement. They range from one-celled organisms, through colonial types, then by slow progress to simple or branched rows of cells (filaments), and finally to elaborate structures attaining sizes and complexities that vie with those of flowering plants. One difficulty in the classification of seaweeds is that the general appearance and size of the plant give almost no clue to the family or genus to which it belongs. The classification depends almost entirely upon the life history, which is especially complex in red algae.

POLYSIPHONIA COLLINSII (above)

Natural size

POLYSIPHONIA HENDRYI (below)

Natural size

Pterosiphonia bipinnata   Black tassel
(Postels and Ruprecht) Falkenberg

Black tassel is a name only partially suitable for this alga. When the branches are matted together on the beach the plant does resemble a long black tassel, but when they are separated the plant is branched into several orders, the uppermost terminating in extremely delicate branchlets, and the whole is fragile and dainty. The alga grows from a rhizoid-like base from Alaska to southern California at mean tide level.

The tassel effect is brought about by the curling back of the branches which causes them to mat together. An average black tassel is 4-8 inches long and ½ inch wide when matted. Individually the branches are thread-like. When the cylindrical, lax branches are separated, one sees they are regularly alternate, the branches arising from segments on the erect shoots and all lying on one plane. In the primary orders the branches are often worn away through entwining around one another. Like other species of *Pterosiphonia*, the black tassel has a tierlike structure with 11-13 cells around a central filament. In the major branches there are three segments between successive branches and the segments are 3-8 times as long as wide; in the smaller branches there are two segments between successive branches and the segments are 2-3 times the breadth.

NOTES ON SEA AND SHORE

It was not until the sun broke through the dense cloud cover over the earth millions and millions of years ago that primitive microorganisms floating in the sea developed chlorophyll. How they did this no one knows, but with the aid of sunlight a reaction was brought about between water and carbon dioxide to build the organic substances needed for life. Thus the first true plants came into being. They then were able to progress from microscopic one-celled algae to large branched forms. Another group of organisms, lacking chlorophyll but needing organic food, found they could make a way of life for themselves by devouring the plants. These are the animals. Ever since those dim ages of the past, all animal life has been dependent upon plant life.

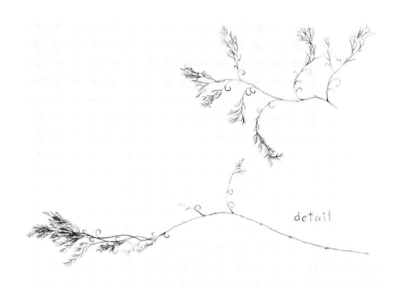

detail

PTEROSIPHONIA BIPINNATA

Natural size

Pterosiphonia dendroidea (Montagne) Falkenberg   **Angel wing**

Pterosiphonia gracilis Kylin   **Baby angel wing**

The extremely delicate angel wings grow in crevices and on protected faces of rocks between mean low- and the -1.5-foot tide levels from Vancouver Island to southern California.

The angel wings are dainty, lacelike, and rose-red to bright red. The species *Pterosiphonia dendroidea* varies from 1 to 3 inches in height, with branches not more than 1/16 inch wide. A number of erect branches arise from a short stipe on one plane and divide into several orders. Each order of branching has a conspicuous compressed axis and featherlike branchlets on each side of it, although the lower part of the branch is often naked.

Pterosiphonia gracilis is about the same height as *P. dendroidea*, but the branches are as narrow as the finest threads. This species, too, is more lacy, fragile, and laxly arranged. In fact, it is one of the most delicate species of seaweed on the Pacific Coast. The branches are on one plane, cylindrical, and irregularly arranged along an indistinct axis. In the final order of branching the branches are in two vertical rows along opposite edges of the branch.

In the angel wing seaweeds the older branches trail along the ground; the younger ones are erect. The trailing branches eventually form rhizoids which act as holdfasts. Both species have a tierlike structure and are made up of a series of microscopic tubes.

NOTES ON SEA AND SHORE

The rock pools in the intertidal belt show an interesting distribution of algae. They harbor a vegetation differing markedly from that on the rocks nearby, although it varies in character at different levels. In the deep pools there is some vertical zonation. The algae in these pools do not dry out, but in the higher pools, exposed for long periods to direct sunlight, the water undergoes extreme temperature variations as well as changes in salt and hydrogen-ion concentrations. Different algae can endure different degrees of acidity or alkalinity. The younger parts of plants are more susceptible than the older parts to pH changes.

PTEROSIPHONIA GRACILIS (left)

Above, twice natural size; below, natural size

PTEROSIPHONIA DENDROIDEA (right)

Twice natural size

Pterochondria woodii (Harvey) Hollenberg   Tassel wing

Like a number of other seaweeds, the tassel wing is epiphytic on other algae. It is usually attached to *Macrocystis, Laminaria, Cystoseira osmundacea,* and a few species of red algae from Vancouver Island to southern California.

The tassel wing is an unmistakable species because it hangs in rather dense masses or tassels from the stipes or branches of the host plant. Although the tassels are in dense masses, the individual thalli are delicate and lacy, usually 4-5 inches long but occasionally 7-8 inches. The seaweed is made up of a freely branched thallus which is alternately and laxly divided into several orders of smaller branches. The hairlike branches are flattened and all lie on one plane. Delicate lacy branchlets, 1/16 inch wide, are arranged at rather infrequent intervals along the sides of the branch. The tips of the branchlets are bluntly forked, the tips turning away from one another. The lower part of the alga is usually a yellowish brown, while the upper part is a deep wine-red. The seaweed is attached to the host by minute rhizoids which penetrate deeply into the host plant. The male and female organs are on separate plants.

NOTES ON SEA AND SHORE

A greater number of epiphytes live on the upper surface of the host plant than on the under side because of the greater intensity of light on the upper side. Because of the response to light there are not so many epiphytes on a host densely covered with branches as there are on less densely branching forms. Depressions in the surface of the host plant also increase the number of epiphytes by providing places for the spores to lodge. An interesting example of this is the large brown alga, *Nereocystis leutkeana.* Early in the summer season, when the stipe is smooth and shiny, it is free of epiphytes, but as the stipe becomes old and wrinkled it has many algae and tiny animals adhering to it. In addition, by the late summer the epiphytes have had a much longer period to settle on the *Nereocystis* than they had had earlier.

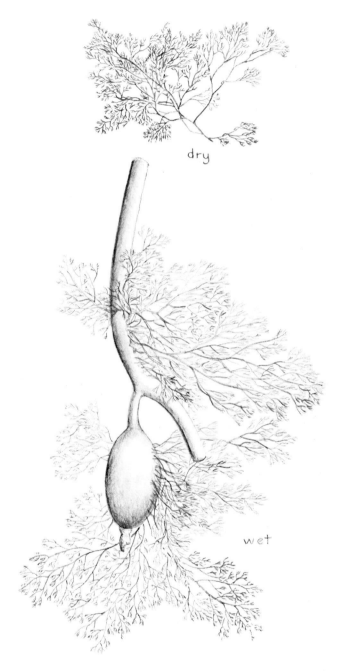

PTEROCHONDRIA WOODII on MACROCYSTIS

Natural size

Laurencia spectabilis Postels and Ruprecht   Sea laurel

The sea laurel grows abundantly on rocks between the mean low- and the -1.5-foot tide levels, from Alaska to southern California.

The striking feature of the sea laurel is the presence of smooth, blunt, flattened, cartilaginous branches which are without midribs, lie on one plane, are free from one another, and are naked on the lower part. The seaweed is usually 5-7 inches in height, occasionally 10 inches, and $1/8$-$1/4$ inch in width. The branches are erect, strong, sometimes solitary, sometimes in clusters. The lower third of the branch is undivided, but toward the top the branch divides alternately to suboppositely. The branches near the bottom are long, and those near the top are short and simple, the top branchlet of each widening at the apex with a slight indentation containing a cell (apical) which initiates the growth. After a certain length of time the apical cell ceases to divide and the apical depression broadens to provide for the swelling of the fertile tips of the male plants. This swelling can be seen at times with the naked eye. The sea laurel is a dull purplish red. Like so many small red seaweeds, the sea laurel has a tiny (1/16 inch), disclike holdfast. In this genus the cell walls are colored blue by a deposit of iodine. Numerous species of *Laurencia* give off an iridescence which is produced by the presence of tannins in the cells.

NOTES ON SEA AND SHORE

The presence or absence of seaweed in an area depends largely upon the physical nature of the substratum. The greatest number of species and individuals live on rocky shores in well-marked communities. The sandy beaches have few seaweeds because the sands are not able to hold and protect the spores which settle on them. Most of the seaweeds found on sandy shores have been torn loose from their holdfasts and washed onto the shore. The rock pools in the intertidal belt also harbor a vegetation different from that on the nearby rocks, and the seaweeds in them vary according to the levels on the shore.

LAURENCIA SPECTABILIS

Natural size

Janczewskia gardneri
Setchell and Guernsey

Parasitic sea laurel

Since this alga is parasitic on the sea laurel *(Laurencia spectabilis)*, its habitat and range are the same as those of its host. Its only hosts appear to be *Laurencia* and *Chondria* in the temperate zones.

The noticeable feature of this seaweed is that it does not look at all like a seaweed. It grows on the host plant in irregularly shaped, cushionlike patches covered with limelike, nodular outgrowths. The cushion is a solidly compact tissue composed of short, thin-walled cells, with a diameter of $1/8$-$1/4$ inch and a height $1/8$ inch, which grow from the tips of cells buried in the pits of *Laurencia*. The cushions are usually remote from one another and are on the flattened older part of the host. The major part of the plant is external to the host, getting its nourishment from the organic matter of the host, which does not, however, appear to be weakened by the parasite. There are few examples of true parasitism in seaweeds. For the most part the host and guest are only superficially attached. The parasitic sea laurel is a neutral yellow-beige or pale pink. The holdfast is a minute, rootlike filament (rhizoid) which penetrates the host by establishing pit connections with its cells.

NOTES ON SEA AND SHORE

From the earliest times man has utilized seaweeds for food, medicine, and fertilizer. Seaweeds have been considered of medicinal value in the Orient since the time of Shen Nung, the father of husbandry and medicine, who lived about 3000 B.C. A poem in the *Chinese Book of Poetry,* written in the time of Confucius, 800-666 B.C., mentions a housewife who cooked seaweed. It appears certain that utilization of seaweed in the Orient was far ahead of its use in the Western world. Nevertheless, the value of Irish moss as a food and of kelps and rockweed as fertilizer was known in Europe long before trade with the Orient began.

JANCZEWSKIA GARDNERI on LAURENCIA SPECTABILIS

Natural size

Rhodomela larix (Turner) C. Agardh   Black pine

The black pine is a seaweed which grows on rocks between the 2- and the 0.5-foot tide levels from Alaska to central California. It is typically an Arctic species that has traveled into warmer waters.

The black pine gets its name from the tufts of short, dark, closely set, spirally arranged branchlets or needles on the branches and stems. These resemble pine needles in texture and stiffness. This seaweed usually reaches a height of 4-7 inches with primary branches about ⅛ inch wide. The primary branches are irregularly arranged around the shoots and become progressively shorter, and the secondary branches are naked, the needles having been worn off along the greater part of their length so that there are only tufts of needles at the tips of the branches. Several erect, cylindrical or slightly compressed shoots arise from a common base. When dry this seaweed is jet black; when alive it is brownish black. The holdfast is a small, thin disc attached to a stone or shell. *Rhodomela larix* is a perennial species whose appearance differs greatly in winter and in summer. In the winter the needles of the frond drop off, leaving the long, lateral branches and the main stem standing stiff and naked. But in the spring and early summer the plant is reclothed in dark brown needles.

NOTES ON SEA AND SHORE

The sea is more homogeneous than the land. For this reason the plants and animals of the sea are more primitive and less diversified than those on land. The seasonal changes in the upper waters of the oceans are less than 5 degrees, and below the 800-foot level there is almost no change in temperature. Moreover, the temperature variation in the sea is only from a high of 96 degrees in the tropics to a low of 29 degrees in polar regions. The medium of water, denser than air, does away with the necessity of rigid stems and trunks in seaweeds. In general, marine algae are not forced to specialize as land plants do or to to adapt themselves to such contrasting habitats as deserts, swamps, jungles, and grassy places.

RHODOMELA LARIX

Natural size

***Odonthalia floccosa* (Esper) Falkenberg   Sharp tooth brush**
***Odonthalia lyallii* (Harvey) J. Agardh   Soft tooth brush**

It is difficult to distinguish the species of *Odonthalia*, but certain characteristics are common to the genus. A typical example is *Odonthalia floccosa*, an abundant perennial species whose major branches, cylindrical to compressed, vary in height from 5-14 inches and from 1/32-1/16 inch in diameter. A plant may have half a dozen stiff branches, with a midrib extending through the greater length, but the midrib becomes lost in a profusion of short branchlets at the tip. The major branches, alternate and widely spaced on opposite sides of the midrib, have short, sharply toothed branchlets running from base to apex. The tips of the branchlets are flattened and recurved, giving a stubby, compact appearance to the plant. In the terminal branchlets the divisions are more numerous on the left side than on the right. The male organs cover the surface of the short, simple branchlets, while the female organs are four-celled and are borne in dense, headlike clusters. When this seaweed is alive it is brownish black; when dead it is a dark brown. The holdfast is a small disc, 1/16 inch in diameter.

Odonthalia lyallii is a much softer and more flexible species than *Odonthalia floccosa*. It does not have a distinct midrib but has branches extending to five or six orders. The over-all length may be 15 inches. The fruiting bodies, both male and female, cluster at the ends of the branchlets, giving a bushy effect.

NOTES ON SEA AND SHORE

For the sea as a whole, changes from night to day, season to season, and age to age are almost unknown, yet the surface of the sea is always changing. These surface variations are brought about by wind, waves, currents, rainfall, and seasons. Of these the greatest changes are seasonal. On the sea as on the land, spring is the time of year for the renewal of life. Of the many factors that contribute to the resurgence of life in the spring, probably the most important is the stirring of the warm bottom waters, which bring minerals to the top for the use of living things.

ODONTHALIA LYALLII (above)

Natural size

ODONTHALIA FLOCCOSA (below)

Natural size

Odonthalia washingtoniensis Kylin   Curry comb

The curry comb is a species of *Odonthalia* found rather commonly from Alaska to Whidbey Island, cast ashore the year around. It is a coarse, brittle species living just below low-tide mark. An average specimen is dark brown or black and reaches a height of 10-15 inches.

The curry comb is striking in appearance, with each branch and branchlet flattened, distinct, separate, and on one plane. The plant arises from a small, disc-shaped holdfast. It has a stipe about 1 inch long, dividing into a number of major branches which in turn branch once or twice. There is no percurrent midrib, although each major branch has a midrib. The edges of the straight, stiff branches are clothed with alternately arranged, widely spaced, spinelike branchlets. The curry comb fruits late in the summer, possibly in August, and bears male and female organs on separate thalli. During the fruiting season the male organs appear on the surface of the short, simple branchlets, especially in the crotch between the terminal branch and the spiny branchlet. The female organs are inflated and spherical. The point of the ultimate branch where the sex organs are attached is always more or less eroded, making it easy for the spores to become attached.

NOTES ON SEA AND SHORE

When the early development of the earth is considered, it is almost certain that plants lived on land before animals did, for only plants were able to bring about conditions that made the habitation of the land possible. Certainly plants helped make soil from crumbling rock and held back the soil that the great rains would have swept away. Little is known of the first land plants, but they must have been closely related to the larger seaweeds that had learned to live in the shallow waters of the coasts. Here they had developed strong stems and rootlike holdfasts to resist the force and pull of the waves and tides. Eventually some such plants found it possible to become permanently separated from the mother ocean.

ODONTHALIA WASHINGTONIENSIS

Upper left, one-half natural size;
others, natural size

Odonthalia kamtschatica (Ruprecht) J. Agardh

Stiff sea brush

The stiff sea brush, like the other species of *Odonthalia*, is found at all seasons of the year from Alaska to Whidbey Island. Since the usual habitat is below low-tide mark, the specimens one finds on the shore have been cast up by the tides. The stiff sea brush, as the name indicates, is coarse and brittle. It is dark brown or black.

This species of *Odonthalia* becomes 8-10 inches high; it has a small, disc-shaped holdfast, and a stipe about 1 inch long. At the upper end of the stipe two cylindrical branches with prominent midribs appear. Arising from the midrib are widely and irregularly spaced alternate branches. The primary branches likewise divide alternately to the third or fourth degree, terminating in branchlets. All the branches are approximately 1/32 inch wide. At intervals along the branches short spine-like growths, about ¼ inch long, appear. Fertilized fruiting bodies are so prominent on the uppermost branches that they made the tips bushy. This species of seaweed seems to fruit late in the summer. Since it is difficult to distinguish the species of *Odonthalia*, there is considerable confusion in the species names of this genus.

NOTES ON SEA AND SHORE

Every marine plant and animal is dependent upon the minerals in the water, and with the increase in minerals which comes about with the stirring of the waters in the spring there is an astonishing increase in the diatoms and other minute plant life. So rapid is their growth that they almost blanket the surface of the sea with a red or brown film. Very soon there is a similar increase in the number of small animals in the plankton. In the spring practically every animal in the sea spawns until the waters "become a vast nursery." These swarms of hungry little creatures floating in the water feed upon the abundant plants. They also devour animals smaller than themselves and in turn are eaten by the larger ones. In the spring migrating fish and whales appear near the shore, intent upon partaking of the rich spring harvest.

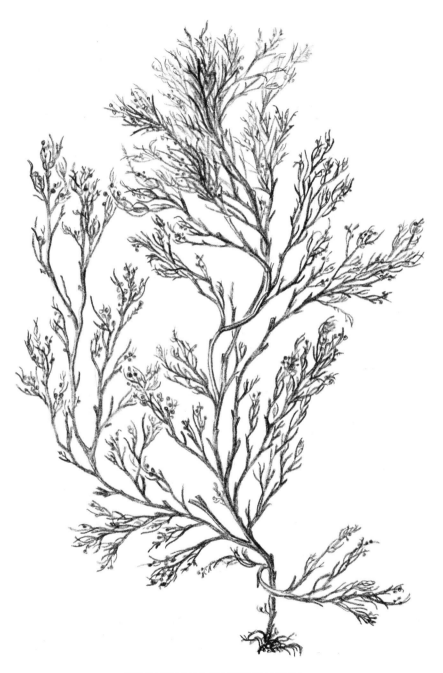

ODONTHALIA KAMTSCHATICA

Natural size

Bibliography

Armstrong, E. Frankland, and L. Mackenzie Miall. *Raw Materials of the Sea.* Brooklyn, N.Y.: Chemical Publishing Company, Inc., 1946.

Burlew, John S. *Algal Culture, from Laboratory to Pilot Plant.* (Carnegie Institution of Washington Publication 600.) Washington, D.C., 1953.

Capt, Lucile. *The Morphology and Life History of Antithamnion.* (Publications of the Puget Sound Biological Station, Vol. VII, No. 35.) Seattle, 1930.

Carson, Rachel. *The Sea Around Us.* New York: Oxford University Press, 1951.

Chapman, V.J. *An Introduction to the Study of Algae.* New York: The Macmillan Press, 1941.

──────. *Seaweeds and Their Uses.* London: Methuen and Company, 1950.

Collins, Frank S. *Green Algae of North America.* New York: G.E. Stechert and Company, 1928.

Dawson, E. Yale. *A Guide to the Literature and Distributions of the Marine Algae of the Pacific Coast of North America.* (Reprinted from *Memmoirs of the Southern California Academy of Science,* Vol. III, No. 1, 1946.)

Fritsch, F.E. *The Structure and Reproduction of the Algae.* Vol. I., New York: The Macmillan Company, 1935; Vol. II, Cambridge, England: The University Press, 1945.

Frye, T.C. *The Age of Pterygophora Californica.* (Publications of the Puget Sound Biological Station, Vol. II, No. 35.) Seattle, 1918.

──────. *Nereocystis Luetkeana.* Chicago: University of Chicago Press. (Reprinted from *Botanical Gazette,* XLII, 143-46, 1906.)

──────, George B. Rigg, W.C. Crandall. *The Size of Kelps on the Pacific Coast of North America.* (Reprinted from *Botanical Gazette,* Vol. XL, No. 6, 1915.)

Gratten, W.H. *British Marine Algae.* London: The Bazaar Office, n.d.

Hartge, Lena Armstrong. *Nereocystis.* (Publications of the Puget Sound Biological Station, Vol. VI, No. 113.) Seattle, 1928.

Harvey, William Henry. *Synopsis of British Seaweeds.* London: Lowell Reeve, 1857.

Hervey, A.B. *Sea Mosses, a Collector's Guide.* New York: Orange Judd Company, 1881.
Kylin, Harald. *The Marine Red Algae in the Vicinity of the Biological Station at Friday Harbor, Washington.* Lund, Sweden: C.W.K. Gleerup, 1925.
Manual of Phycology, ed. Gilbert W. Smith. Waltham, Mass.: Published by the Chronica Botanica Company, 1951.
Muenscher, Walter C. *A Key to the Phaeophyceae of Puget Sound.* (Publications of the Puget Sound Biological Station, Vol. I, No. 25.) Seattle, 1917.
Newton, Lily. *Seaweed Utilisation.* London: Sampson Low, 1951.
──────. *A Handbook of the British Seaweeds.* London: The Trustees of the British Museum, 1931.
Rigg, George B. *Seasonal Development of Bladder Kelp.* (Publications of the Puget Sound Biological Station, Vol. I, No. 27.) Seattle, 1917.
──────, and Robert C. Miller. "Intertidal Plant and Animal Zonation at Neah Bay, Washington," *Proceedings of the California Academy of Sciences* (San Francisco), Vol. XXVI, No. 10 (1949), pp. 323-51.
Setchell, William Albert, and Nathaniel Lyon Gardner. *Algae of Northwestern America.* Berkeley: University of California Press, 1903.
──────. *The Marine Algae of the Pacific Coast of North America,* Part III. Berkeley: University of California Press, 1925.
Smith, Gilbert M. *Marine Algae of the Monterey Peninsula.* Stanford: Stanford University Press, 1951.
Sverdrup, H.U., Martin W. Johnson, Richard H. Fleming. *The Oceans.* New York: Prentice Hall, Inc., 1946.
Tiffany, Lewis Hanford. *Algae, the Grass of Many Waters.* Baltimore: Charles C. Thomas, 1938.
Tilden, Josephine E. *The Algae and Their Life Relations.* Minneapolis: University of Minnesota Press, 1944.
Tressler, Donald, and James McW. Lemon. *Marine Products of Commerce.* New York: Reinhold Publishing Corporation, 1951.

Scientific Name Index

Agardhiella coulteri, 104
Agarum fimbriatum, 54
Ahnfeltia gigartinoides, 112
Ahnfeltia plicata, 112
Alaria valida, 66
Antithamnion pacificum, 128
Antithamnion uncinatum, 128

Bossea sp., 92
Bryopsis corticulans, 16

Calliarthron sp., 94
Callithamnion pikeanum, 130
Callophyllis edentata, 100
Ceramium pacificum, 134
Cladophora trichotoma, 12
Codium fragile, 18
Codium setchellii, 20
Coilodesme californica, 38
Colpomenia sinuosa, 36
Constantinea simplex, 84
Corallina chilensis, 90
Corallina gracilis, 90
Costaria costata, 52
Cryptopleura ruprechtiana, 150
Cyamathere triplicata, 50
Cystophyllum geminatum, 72
Cystoseira osmundacea, 74

Dasyopsis plumosa, 152
Delesseria decipiens, 142
Desmarestia aculeata, 26
Desmarestia intermedia, 28
Desmarestia munda, 30

Egregia menziesii, 68
Endocladia muricata, 86
Enteromorpha compressa, 2
Enteromorpha intestinalis, 4
Enteromorpha plumosa, 6
Erythroglossum intermedium, 144

Farlowia mollis, 82
Fucus furcatus, 70

Gastroclonium coulteri, 126
Gigartina sp., 114
Gigartina exasperata, 116
Gracilaria confervoides, 110
Grateloupia pinnata, 96
Griffithsia pacifica, 132

Halosaccion glandiforme, 120
Hedophyllum sessile, 56
Heterochordaria abietina, 24
Hymenena flabelligera, 148

Iridophycus sp., 118

Janczewskia gardneri, 166

Laminaria andersonii, 46
Laminaria bullata, 44
Laminaria platymeris, 40
Laminaria saccharina, 42
Laurencia spectabilis, 164
Lithothamnium sp., 88

Macrocystis integrifolia, 62

Macrocystis pyrifera, 62
Microcladia borealis, 138
Microcladia coulteri, 136

Nereocystis luetkeana, 58

Odonthalia floccosa, 170
Odonthalia kamtschatica, 174
Odonthalia lyallii, 170
Odonthalia washingtoniensis, 172

Pleurophycus gardneri, 48
Plocamium pacificum, 108
Polyneura latissima, 146
Polysiphonia collinsii, 156
Polysiphonia hendryi, 156
Polysiphonia pacifica, 154
Porphyra lanceolata, 80
Porphyra naiadum, 76
Porphyra perforata, 78
Postelsia palmaeformis, 60

Prionitis lyallii, 98
Pterochondria woodii, 162
Pterosiphonia bipinnata, 158
Pterosiphonia dendroidea, 160
Pterosiphonia gracilis, 160
Pterygophora californica, 64
Ptilota filicina, 140
Punctaria latifolia, 32

Ralfsia pacifica, 22
Rhodomela larix, 168
Rhodymenia palmata, 122
Rhodymenia pertusa, 124

Sarcodiotheca furcata, 106
Schizymenia pacifica, 102
Scytosiphon lomentaria, 34
Spongomorpha coalita, 14

Ulva lactuca, 8
Ulva linza, 10

Common Name Index

Ahnfelt's seaweed, bushy, 112
Ahnfelt's seaweed, loose, 112
Angel wing, 160

Baby angel wing, 160
Baron Delessert, 142
Bead coral, 94
Beauty bush, 130
Black pine, 168
Black tassel, 158
Bladder leaf, 72
Blister wrack, 44
Brown sieve, 32
Bull kelp, 58

Color changer, crisp, 26
Color changer, loose, 28
Color changer, wide branch, 30
Coulter's seaweed, 104
Crisscross network, 146
Cup and saucer, 84
Curry comb, 172

Dainty leaf, 144
Dulse, 122

Farlow seaweed, 82
Feather boa, 68
Fir needle, 24

Graceful coral, 90
Grapestone, 114
Green ball, 12
Green confetti, 2

Green rope, 14
Green string lettuce, 10
Griffith seaweed, 132

Hidden rib, 150
Honey ware, 66
Hooked rope, 128
Hooked skein, 128

Iridescent seaweed, 118

Leaf coral, 92
Link confetti, 4
Lyall's seaweed, 98

Nail brush, 86

Oyster thief, 36

Parasitic sea laurel, 166
Perennial kelp, giant, 62
Perennial kelp, small, 62
Plumed chenille, 152
Pocket thief, 36
Pointed lynx, 96
Polly Collins, 156
Polly Hendry, 156
Polly Pacific, 154
Pompon, 64
Popping wrack, 70
Pottery seaweed, 134

Red eyelet silk, 124
Red fringe, 76

181

Red jabot laver, 80
Red kale, 122
Red laver, 78
Red rock crust, 88
Red sea fan, 100
Red serving fork, 106
Red wing, 140
Ribbon kelp, 58
Rockweed, 70
Ruche, 150

Sea belly, 126
Sea brush, soft, 170
Sea brush, stiff, 174
Sea cabbage, 56
Sea colander, 54
Sea comb, 108
Sea fern, 16
Sea girdle, 40
Sea lace, coarse, 138
Sea lace, delicate, 136
Sea laurel, 164
Sea lettuce, 8
Sea palm, 60

Sea rose, 102
Sea sac, 120
Sea spatula, 48
Sea staghorn, 18
Seersucker, 52
Sewing thread, 110
Sharp tooth brush, 170
Silk confetti, 6
Split whip wrack, 46
Spongy cushion, 20
Stick bag, 38
Sugar wrack, 42

Tangle, 40
Tar spot, 22
Tassel wing, 162
Tide pool coral, 90
Triple rib, 50
Turkish towel, 116

Veined fan, 148

Whip tube, 34
Wing kelp, 66
Woody chain bladder, 74